Quick Calculus

Second Edition
A Self-Teaching Guide

Daniel Kleppner, Ph.D.
Lester Wolfe Professor of Physics
Massachusetts Institute of Technology

Norman Ramsey, Ph.D.
Higgins Professor of Physics
Harvard University
Nobel Prize for Physics 1989

A Wiley Press Book
John Wiley & Sons, Inc.
New York · Chichester · Brisbane · Toronto · Singapore

Publisher: Stephen Kippur
Editor: Elizabeth G. Perry
Managing Editor: Katherine Schowalter
Composition & Make-up: Cobb-Dunlop Publisher Services

Library of Congress Cataloging-in-Publication Data

Kleppner, Daniel.
Quick calculus.

Includes indexes.
1. Calculus—Programmed instruction.
I. Ramsey, Norman, 1915– . II. Title.
QA303.K665 1985 515'.07'7 85-12349
ISBN 0-471-82722-3

Printed in the United States of America

10 9 8 7

Preface

Before you plunge into *Quick Calculus,* perhaps we ought to tell you what it is supposed to do. *Quick Calculus* should teach you the elementary techniques of differential and integral calculus with a minimum of wasted effort on your part; it is designed for you to study by yourself. Since the best way for anyone to learn calculus is to work problems, we have included many problems in this book. You will always see the solution to your problem as soon as you have finished it, and what you do next will depend on your answer. A correct answer generally sends you to new material, while an incorrect answer sends you to further explanations and perhaps another problem.

We hope that this book will be useful to many different people. The idea for it grew out of the problem of teaching college freshmen enough calculus so that they could start physics without waiting for a calculus course in college. However, it soon became apparent that the book would be useful in many other ways. For instance, both graduate and undergraduate students in economics, business, medicine, and the social sciences need to use some elementary calculus. Many of these students have never taken calculus, or want to review the course they did take; they should be able to put this book to good use. Ambitious high school students who want to get a head start on their college studies should find *Quick Calculus* just the thing. Unlike most calculus texts, it emphasizes technique and application rather than rigorous theories and is therefore particularly suited for introducing the subject. Beginning calculus students who want a different and simpler view of the subject should find the book helpful either for self-instruction or for classroom use. We particularly hope that this book will be of use to those people who simply want to learn calculus for the fun of it.

Because of the variety of backgrounds of those who will use this book, we start with a review of some parts of algebra and trigonometry which are useful in elementary calculus. If you remember your high school preparation in these subjects, you will sail through this material in little time, whereas if you have had little math, or have long been away from math, you will want to spend more time on this review. As you will see, one of the virtues of the book is its flexibility—the time you spend on each portion depends on your particular needs. We hope that this will save you time so that you will find the book's title appropriate.

Daniel Kleppner
Norman Ramsey

Cambridge, Massachusetts

Preface to the Second Edition

The hope expressed in the preface to the first edition that *Quick Calculus* would be useful to many different people has been fulfilled, for over a quarter of a million copies have been put to use. The major change in the second edition is in the treatment of integration. Chapter 3 has been completely rewritten; much of the material has been simplified, and a new topic has been added—numerical integration. In addition, numerical exercises to be worked with a hand-held calculator have been introduced throughout the book. These are not an essential part of the text, but we hope that readers who have a calculator will find them interesting and useful. Many minor improvements have been made, and the references have been updated.

Daniel Kleppner
Norman Ramsey

Cambridge, Massachusetts

Contents

More than 100 Wiley Self-Teaching Guides teach practical skills on everything from accounting to astronomy, from microcomputers to mathematics.

STGs FOR MATHEMATICS

Accounting Essentials, 2nd Edition, Margolis & Harmon
Finite Mathematics, Rothenberg
Geometry and Trigonometry for Calculus, Selby
Linear Algebra with Computer Applications, Rothenberg
Management Accounting, Madden
Math Shortcuts, Locke
Math Skills for the Sciences, Pearson
Practical Algebra, Selby
Quick Algebra Review, Selby
Quick Arithmetic, 2nd Edition, Carman & Carman
Quick Calculus, 2nd Edition, Kleppner & Ramsey
Statistics, 3rd Edition, Koosis

Look for these and other Self-Teaching Guides at your favorite bookstore!

CHAPTER ONE
A Few Preliminaries

In this chapter the plan of the book is explained, and some elementary mathematical concepts are reviewed. By the end of the chapter you will be familiar with:

- The definition of a mathematical function
- Graphs of functions
- The properties of the most widely used functions: linear and quadratic functions, trigonometric functions, exponentials, and logarithms

Some problems in *Quick Calculus* require the use of a scientific calculator—a calculator that provides values for trigonometric functions and logarithms. However, these problems, which are clearly marked, are optional. You can skip them and master the text without a calculator, although working the numerical problems will help to increase your insight.

Getting Started

1

In spite of its formidable name, calculus is not a particularly difficult subject. Of course you won't become a master in it overnight, but with diligence you can learn its basic ideas fairly quickly.

This manual will get you started in calculus. After working through it, you ought to be able to handle many problems and you should be prepared to learn more elaborate techniques if you need them. But remember that the important word is *working,* though we hope you find that much of the work is fun.

(continued)

Most of your work will be answering questions and doing problems. The particular route you follow will depend on your answers. Your reward for doing a problem correctly is to go straight on to new material. On the other hand, if you make an error, the solution will usually be explained and you will get additional problems to see whether you have caught on. In any case, you will always be able to check your answers immediately after doing a problem.

Many of the problems have multiple choice answers. The possible choices are grouped like this: [a | b | c | d]. Choose an answer by circling your choice. The correct answer can be found at the bottom of the next left-hand page. Some questions must be answered with written words. Space for these is indicated by a blank, and you will be referred to another frame for the correct answer.

If you get the right answer but feel you need more practice, simply follow the directions for the wrong answer. There is no premium for doing this book in record time.

Go on to frame **2**.

2

In case you want to know what's ahead, here is a brief outline of the book: this first chapter is a review which will be useful later on; Chapter 2 is on differential calculus; and Chapter 3 covers integral calculus. Chapter 4 contains a concise outline of all the earlier work. There are two appendixes—one giving formal proofs of a number of relations we use in the book and the other discussing some supplementary topics. In addition, there is a list of extra problems, with answers, and a section of tables you may find useful.

A word of caution about the next few frames. Since we must start with some definitions, the first section has to be somewhat more formal than most other parts of the book.

First we review the definition of a function. If you are already familiar with this, and with the idea of independent and dependent variables, you should skip to frame **14**. (In fact, in this chapter there is ample opportunity for skipping if you already know the material. On the other hand, some of the material may be new to you, and a little time spent on review can be a good thing.)

Go to **3**.

Functions

3 ———

The definition of a function makes use of the idea of a *set*. Do you know what a set is? If so, go to **4**. If not, read on.

A *set* is a collection of objects—not necessarily material objects—described in such a way that we have no doubt as to whether a particular object does or does not belong to it. A set may be described by listing its elements. Example: the set of numbers, 23, 7, 5, 10. Another example: Mars, Rome, and France.

We can also describe a set by a rule, for example, all the even positive integers (this set contains an infinite number of objects). Another set defined by a rule is the set of all planets in our solar system.

A particularly useful set is the set of all real numbers, which includes all numbers such as $5, -4, 0, \frac{1}{2}, -3.482, \sqrt{2}$. The set of real numbers does *not* include quantities involving the square root of negative numbers (such quantities are called *complex numbers;* in this book we will be concerned only with real numbers).

The mathematical use of the word "set" is similar to the use of the same word in ordinary conversation, as "a set of golf clubs."

Go to **4**.

4 ———

In the blank below, list the elements of the set which consists of all the odd integers between -10 and $+10$.

____

Go to **5** for the correct answer.

5 ———

Here are the elements of the set of all odd integers between -10 and $+10$:

$$-9, -7, -3, -1, 1, 3, 5, 7, 9.$$

Go to **6**.

6

Now we are ready to talk about functions. Here is the definition.

A *function* is a rule that assigns to each element in a set A one and only one element in a set B.

The rule can be specified by a mathematical formula such as $y = x^2$, or by tables of associated numbers, for instance, the temperature at each hour of the day. If x is one of the elements in set A, then the element in set B that the function f associates with x is denoted by the symbol $f(x)$. [This symbol $f(x)$ is the value of f at x. It is usually read as "f of x."]

The set A is called the *domain* of the function.

The set B of all possible values of $f(x)$ as x varies over the domain is called the *range* of the function.

In general, A and B need not be restricted to sets of real numbers. However, as mentioned in frame **3**, in this book we will be concerned only with real numbers.

Go to **7**.

7

For example, for the function $f(x) = x^2$, with the domain being all real numbers, the range is

______________________.

Go to **8**.

8

The range is *all nonnegative real numbers*. For an explanation, go to **9**. Otherwise,

Skip to **10**.

9

Recall that the product of two negative numbers is positive. Thus for any real value of x, positive or negative, x^2 is positive. When x is 0, x^2 is also 0. Therefore, the range of $f(x) = x^2$ is all nonnegative numbers.

Go to **10**.

10

Our chief interest will be in rules for evaluating functions defined by formulas. If the domain is not specified, it will be understood that the domain is the set of all real numbers for which the formula produces a real value, and for which it makes sense. For instance,

(a) $f(x) = \sqrt{x}$ Range = ____________________.

(b) $f(x) = \dfrac{1}{x}$ Range = ____________________.

Go to **11**.

11

$f(x)$ is real for x nonnegative; so the answer to (a) is all nonnegative real numbers.

$1/x$ is defined for all values of x except zero; so the range in (b) is all real numbers except zero.

Go to **12**.

12

When a function is defined by a formula such as $f(x) = ax^3 + b$, x is called the *independent variable* and $f(x)$ is called the *dependent variable*. One advantage of this notation is that the value of the dependent variable, say for $x = 3$, can be indicated by $f(3)$.

Often, however, a single letter is used to represent the dependent variable, as in

$$y = f(x).$$

Here x is the independent variable and y is the dependent variable.

Go to **13**.

13

In mathematics the symbol x frequently represents an independent variable, f often represents the function, and $y = f(x)$ usually denotes the dependent variable. However, any other symbols may be used for the function, the independent variable, and the dependent variable. For

(continued)

example, we might have $z = H(r)$ which is read as "z equals H of r." Here r is the independent variable, z is the dependent variable, H is the function.

Now that we know what a function means, let's move along to a discussion of graphs.

Go to **14**.

Graphs

14

If you know how to plot graphs of functions, you can skip to frame **19**. Otherwise,

Go to **15**.

15

A convenient way to represent a function defined by $y = f(x)$ is to plot a graph. We start by constructing coordinate axes. First we construct a pair of mutually perpendicular intersecting lines, one horizontal, the other vertical. The horizontal line is often called the *x-axis*, and the vertical line the *y-axis*. The point of intersection is the *origin*, and the axes together are called the *coordinate axes*.

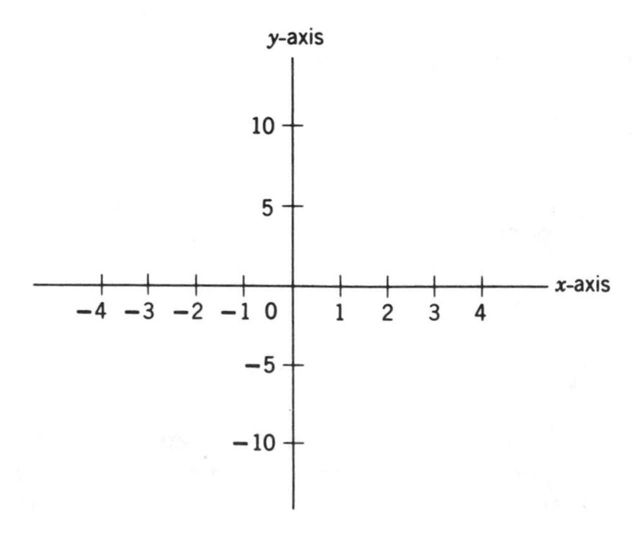

Next we select a convenient unit of length and, starting from the origin, mark off a number scale on the x-axis, positive to the right and negative to the left. In the same way we mark off a scale along the y-axis with positive numbers going upward and negative downward. The scale of the y-axis does not need to be the same as that for the x-axis (as in the drawing). In fact, y and x can have different units, such as distance and time.

Go to **16**.

16

We can represent one specific pair of values associated by the function in the following way: Let a represent some particular value for the independent variable x, and let b indicate the corresponding value of $y = f(x)$. Thus, $b = f(a)$.

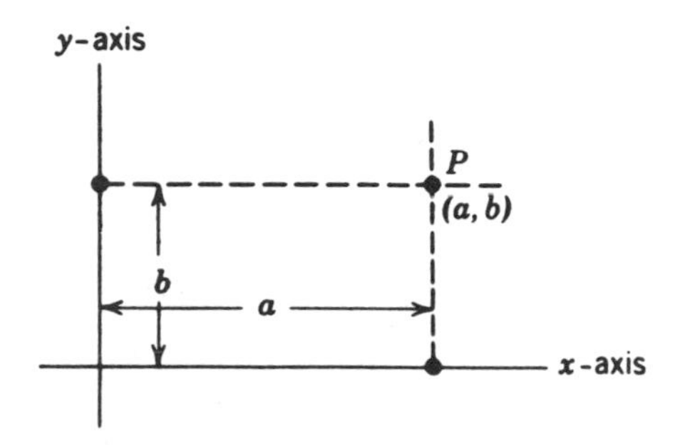

We now draw a line parallel to the y-axis at distance a from that axis, and another line parallel to the x-axis at distance b. The point P at which these two lines intersect is designated by the pair of values (a, b) for x and y, respectively.

The number a is called the x-coordinate of P, and the number b is called the y-coordinate of P. (Sometimes the x-coordinate is called the *abscissa*, and the y-coordinate is called the *ordinate*.) In the designation of a typical point by the notation (a, b) we will always designate the x-coordinate first and the y-coordinate second.

As a review of this terminology, encircle the correct answers below. For the point $(5, -3)$:

x-coordinate: [−5 | −3 | 3 | 5]

y-coordinate: [−5 | −3 | 3 | 5]

(continued)

(Remember that the answers to multiple choice questions are ordinarily given at the bottom of the next left-hand page. Always check your answers before continuing.)

Go to **17**.

17

The most direct way to plot the graph of a function $y = f(x)$ is to make a table of reasonably spaced values of x and of the corresponding values of $y = f(x)$. Then each pair of values (x, y) can be represented by a point as in the previous frame. A graph of the function is obtained by connecting the points with a smooth curve. Of course, the points on the curve may be only approximate. If we want an accurate plot, we just have to be very careful and use many points. (On the other hand, crude plots are pretty good for many purposes.)

Go to **18**.

18

As an example, here is a plot of the function $y = 3x^2$. A table of values of x and y is shown and these points are indicated on the graph.

x	y
−3	27
−2	12
−1	3
0	0
1	3
2	12
3	27

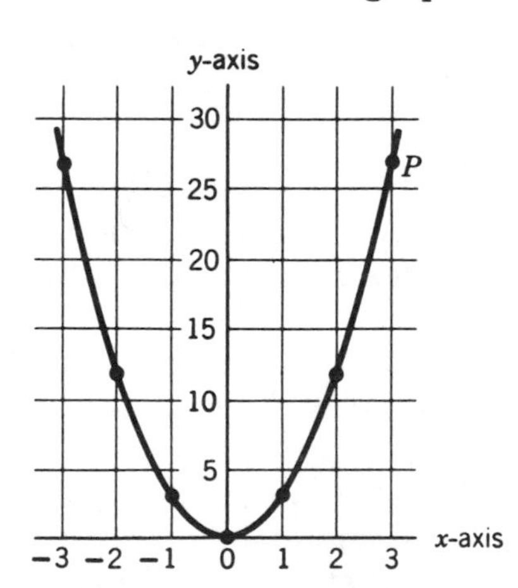

To test yourself, encircle the pair of coordinates that corresponds to the point P indicated in the figure.

[(3, 27) | (27, 3) | none of these]

Answers: (16) 5, −3

Check your answer. If incorrect, study frame **16** once again and then go to **19**. If correct,

Go on to **19**.

19

Here is a rather special function. It is called a *constant function* and assigns a single fixed number c to every value of the independent variable, x. Hence, $f(x) = c$.

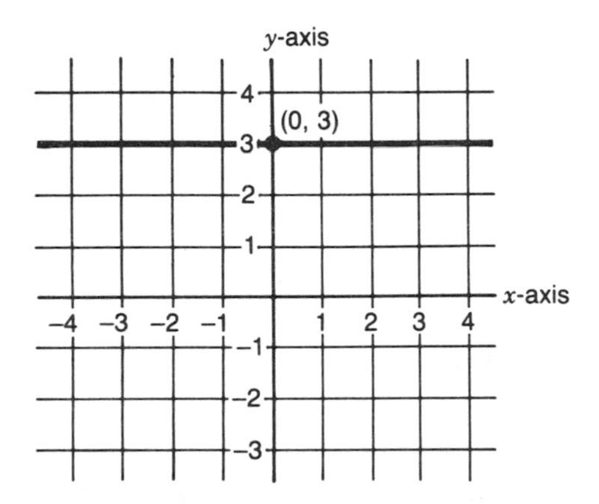

This is a peculiar function since the value of the dependent variable is the same for all values of the independent variable. Nevertheless, the relation $f(x) = c$ assigns exactly one value of $f(x)$ to each value of x as required in the definition of a function. All the values of $f(x)$ happen to be the same.

Try to convince yourself that the graph of the constant function $y = f(x) = 3$ is a straight line parallel to the x-axis passing through the point (0, 3) as shown in the figure.

Go to **20**.

20

Another simple function is the *absolute value function*. The absolute value of x is indicated by the symbols $|x|$. The absolute value of a number x determines the size or magnitude of the number without regard to its sign. For example,

$$|-3| = |3| = 3.$$

(continued)

Now we will define $|x|$ in a general way. But first we should recall the inequality symbols:

$a > b$ means a is greater than b.
$a \geqslant b$ means a is greater than or equal to b.
$a < b$ means a is less than b.
$a \leqslant b$ means a is less than or equal to b.

With this notation we can define the absolute value function, $|x|$, by the following two rules:

$$|x| = \begin{cases} x & \text{if } x \geqslant 0, \\ -x & \text{if } x < 0. \end{cases}$$

Go to **21**.

21

A good way to show the behavior of a function is to plot its graph. Therefore, as an exercise, plot a graph of the function $y = |x|$ in the accompanying figure.

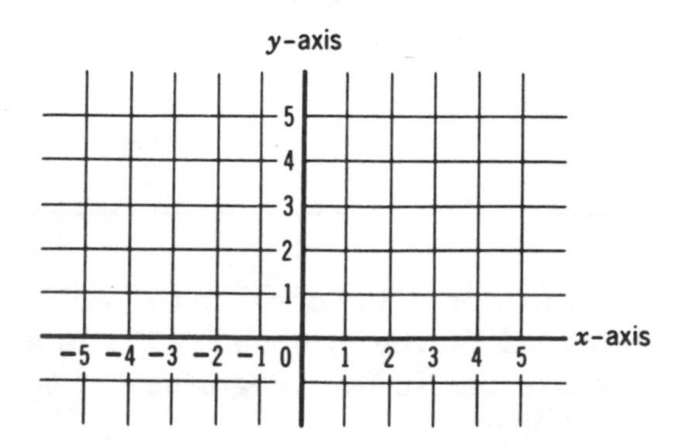

To check your answer, go to **22**.

Answer: (18) (3, 27)

22

The correct graph for $|x|$ is

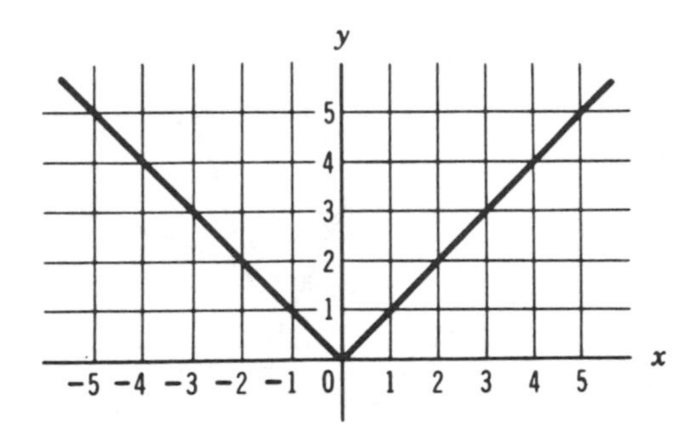

This can be seen by preparing a table of x and y values as follows:

x	$y = \|x\|$
−4	+4
−2	+2
0	0
+2	+2
+4	+4

These points may be plotted as in frames **16** and **18** and the lines drawn with the results in the above figure.

With this introduction on functions and graphs, we are now going to take a quick look at some elementary functions which are important. You should become familiar with them.

These functions are linear, quadratic, trigonometric, exponential, and logarithmic functions.

Go to 23.

Linear and Quadratic Functions

23

A function defined by an equation in the form $y = mx + b$, where m and b are constants, is called a *linear* function because its graph is a straight line. This is a simple and useful function, and you should really become familiar with it.

(continued)

Here is an example: Encircle the letter which identifies the graph of

$$y = 3x - 3. \qquad [A \mid B \mid C]$$

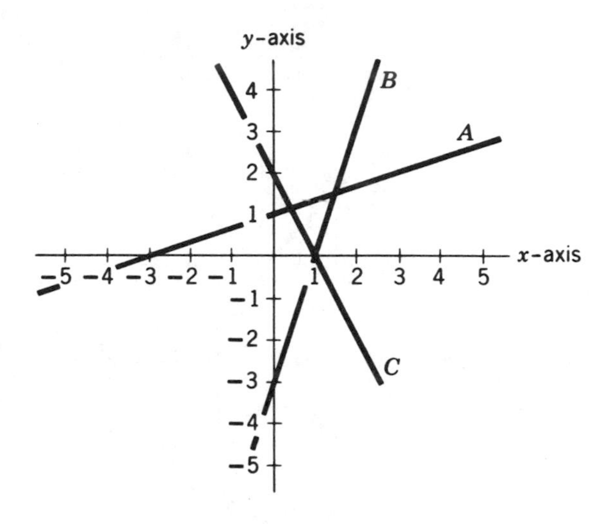

The correct answer is on the bottom of the next page. If you missed this or if you do not feel entirely sure of the answer, go to **24**.

Otherwise, go to **25**.

24

You were given the function $y = 3x - 3$. The table below gives a few values of x and y.

x	y
−2	−9
−1	−6
0	−3
1	0
2	3

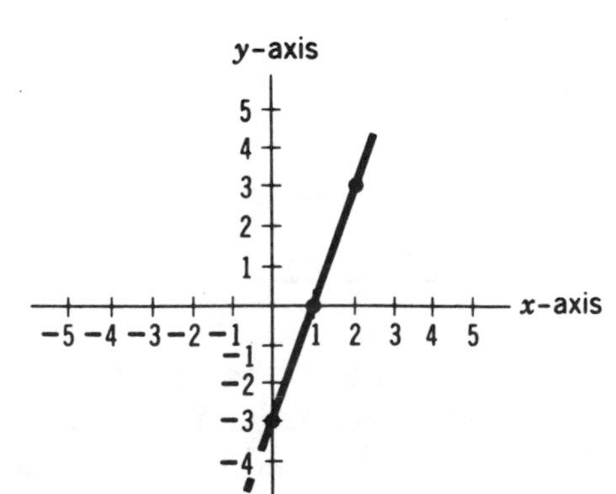

A few of these points are shown on the graph, and a straight line has been drawn through them. This is line B of the figure in frame **23**.

Go to **25**.

25

Here is the graph of a typical linear equation. Let us take any two different points on the line, (x_2, y_2) and (x_1, y_1). We define the slope of the line in the following way:

$$\text{Slope} = \frac{y_2 - y_1}{x_2 - x_1}.$$

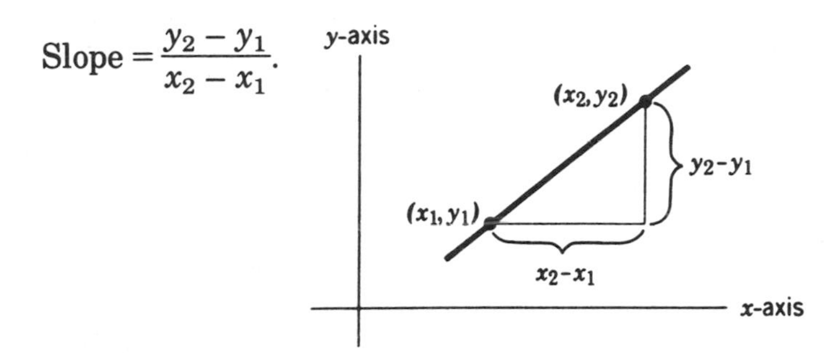

The idea of slope will be very important in our later work, so let's spend a little time learning more about it.

Go to **26**.

26

If the x and y scales are the same, as in the figure, then the slope is the ratio of *vertical* distance to *horizontal* distance as we go from one point on the line to another, providing we take the sign of each line segment as in the equation of frame **25**. If the line is vertical, the slope is infinite (or, more strictly, undefined). It should be clear that the slope is the same for any pair of two separate points on the line.

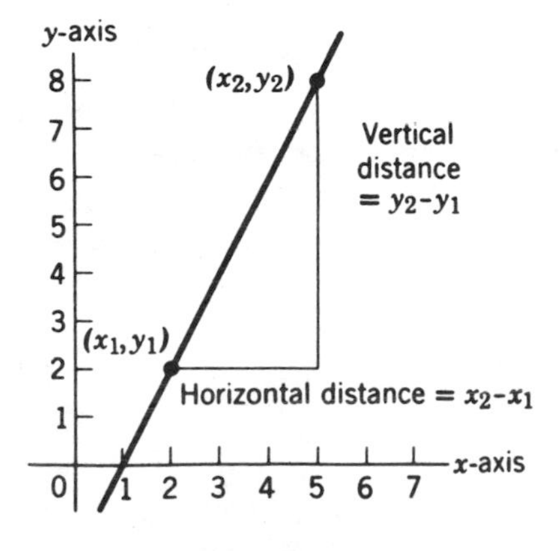

Go to **27**.

27

If the vertical and horizontal scales are not the same, the slope is still defined by

$$\text{Slope} = \frac{\text{vertical distance}}{\text{horizontal distance}},$$

but now the distance is measured using the appropriate scale. For instance, the two figures below may look similar, but the slopes are quite different. In the first figure the x and y scales are identical, and the slope is ½. In the second figure the y scale has been changed by a factor of 100, and the slope is 50.

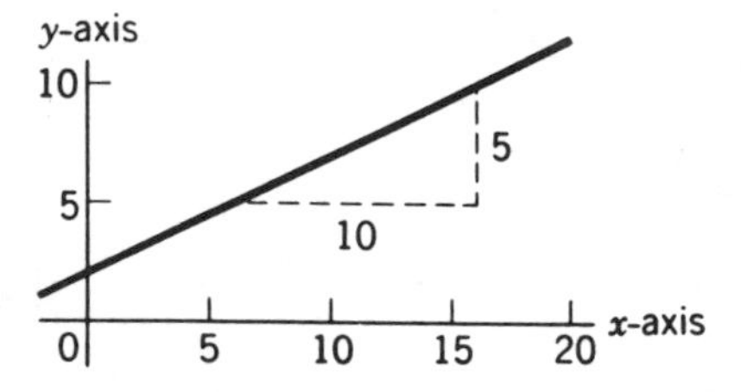

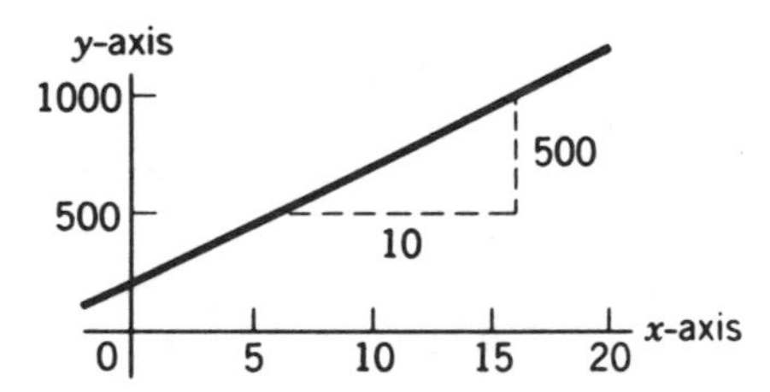

Since the slope is the ratio of two lengths, the slope is a pure number if the lengths are pure numbers. However, if the variables have different dimensions, the slope will also have a dimension.

Below is a plot of the distance traveled by a car vs. the amount of gasoline consumed.

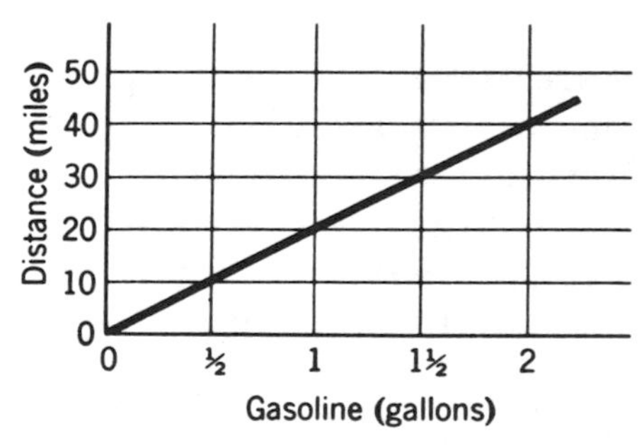

Here the slope has the dimension or unit of miles/gallon (or miles per gallon). What is the slope of the line shown?

Slope = [10 | 20 | 30 | 40] miles/gallon

If right, go to **29**.
Otherwise, go to **28**.

Answer: **(23)** *B*

28

To evaluate the slope, let us find the coordinates of two points on the line.

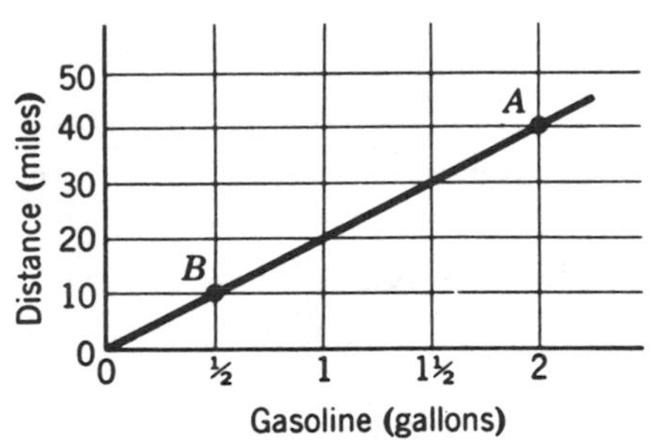

For instance, A has the coordinates (2 gallons, 40 miles) and B has the coordinates (½ gallon, 10 miles). Therefore, the slope is

$$\frac{(40-10)\text{ miles}}{(2-\frac{1}{2})\text{ gallons}} = \frac{30\text{ miles}}{\frac{3}{2}\text{ gallons}} = 20\text{ miles per gallon.}$$

Of course, we would have obtained the same value for the slope no matter which two points we used, since the ratio of vertical distance to horizontal distance is the same everywhere.

Go to **29**.

29

Here is another way to find the slope of a straight line if its equation is given. If the linear equation is in the form $y = mx + b$, then the slope is given by

$$\text{Slope} = \frac{y_2 - y_1}{x_2 - x_1}.$$

Substituting in the above expression for y, we have

$$\text{Slope} = \frac{(mx_2 + b) - (mx_1 + b)}{x_2 - x_1} = \frac{mx_2 - mx_1}{x_2 - x_1} = \frac{m(x_2 - x_1)}{x_2 - x_1} = m.$$

What is the slope of $y = 7x - 5$?

[5/7 | 7/5 | −5 | −7 | 5 | 7]

If right, go to **31**.
Otherwise, go to **30**.

30

The equation $y = 7x - 5$ can be written in the form $y = mx + b$ if $m = 7$ and $b = -5$. Since slope $= m$, the line given has a slope of 7.

Go to **31**.

31

The slope of a line can be positive (greater than 0), negative (less than 0), or 0. An example of each is shown graphically below.

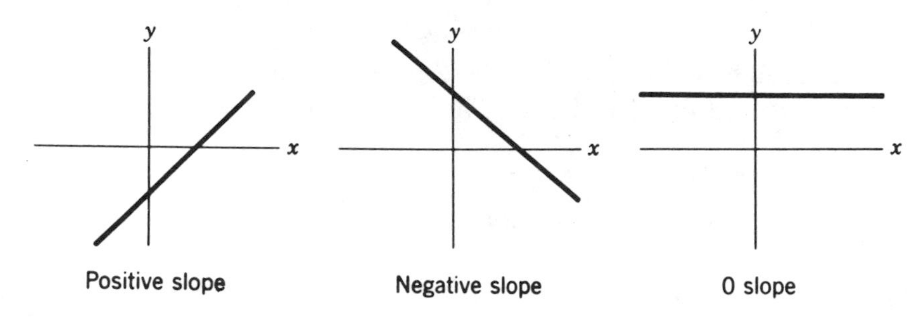

Positive slope　　Negative slope　　0 slope

Note how a line with positive slope arises in going from left to right, while a line with negative slope falls in going from left to right. (It was pointed out in frame **26** that the slope of a vertical line is not defined.)

Indicate whether the slope of the graph of each of the following equations is positive, negative, or zero by encircling your choice.

	Equation	Slope
1.	$y = 2x - 5$	[+ | – | 0]
2.	$y = -3x$	[+ | – | 0]
3.	$p = q - 2$	[+ | – | 0]
4.	$y = 4$	[+ | – | 0]

If all right, go to **33**.
If you made any mistakes, go to **32**.

32

Here are the explanations to the questions in frame **31**.

In frame **29** we saw that for a linear equation in the form $y = mx + b$ the slope is m.

Answer: (27) 20 miles/gallon　　**(29)** 7

1. $y = 2x - 5$. Here $m = 2$ and the slope is 2. Clearly this is a positive number. See Figure 1 below.
2. $y = -3x$. Here $m = -3$. The slope is -3, which is negative. See Figure 2 below.
3. $p = q - 2$. In this equation the variables are p and q, rather than y and x. Written in the form $p = mq + b$, it is evident that $m = 1$, which is positive. See Figure 3 below.
4. $y = 4$. This is an example of a constant function. Here $m = 0$, $b = 4$, and the slope is 0. See Figure 4 below.

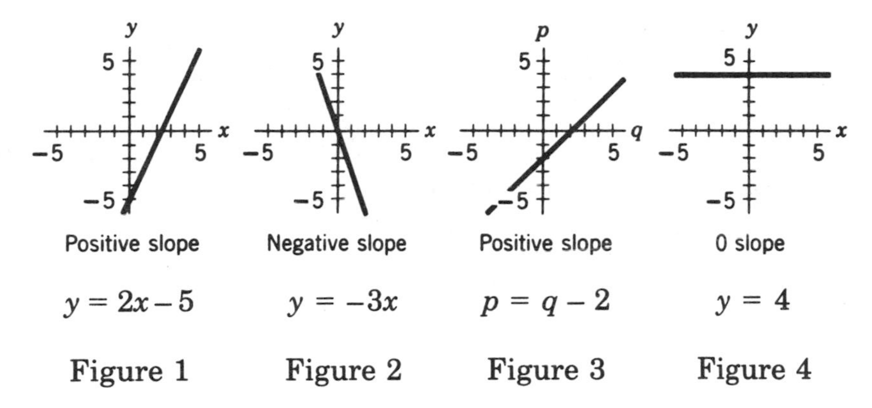

Positive slope	Negative slope	Positive slope	0 slope
$y = 2x - 5$	$y = -3x$	$p = q - 2$	$y = 4$
Figure 1	Figure 2	Figure 3	Figure 4

Go to **33**.

33

Here is an example of a linear equation in which the slope has a familiar meaning. The graph below shows the position S on a straight road of a car at different times. The position $S = 0$ means the car is at the starting point.

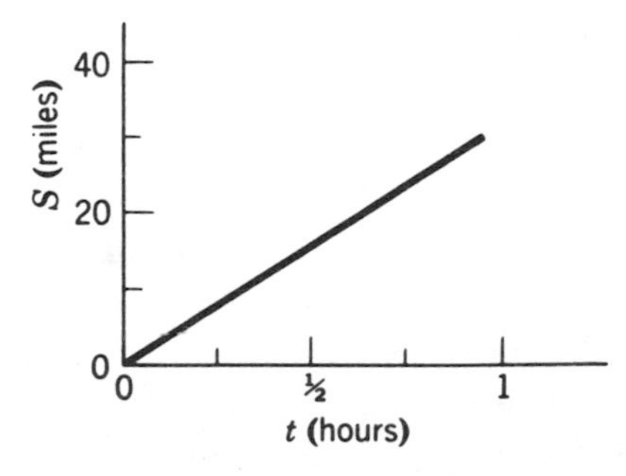

Try to guess the correct word to fill in the blank below:

The slope of the line has the same value as the car's ________.

To see the correct answer, go to **34**.

34

The slope of the line has the same value as the car's *velocity* (or its *speed*).

The slope is given by the ratio of the distance traveled to the time required. But, by definition, the velocity is also the distance traveled divided by the time. Thus the value of the slope of the line is equal to the velocity.

Go to **35**.

35

Now let's look at another type of equation. An equation in the form $y = ax^2 + bx + c$, where a, b, and c are constants, is called a *quadratic* equation and its graph is called a *parabola*. Two typical parabolas are shown in the figure.

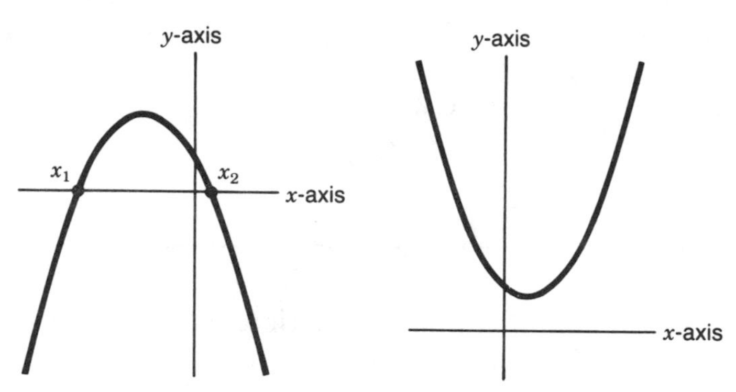

Go to **36**.

36

The values of x at $y = 0$, shown by x_1 and x_2 in the figure on the left in frame **35** correspond to values of x which satisfy $ax^2 + bx + c = 0$ and are called the *roots* of the equation. Not all quadratic equations have real roots. (For example, the curve on the right represents an equation with no real value of x when $y = 0$.)

Although you will not need to find the roots of any quadratic equation later in this book, you may want to know the formula anyway. If you would like to see a discussion of this, go to frame **37**.

Otherwise, skip to frame **39**.

Answers: (31) +, –, +, 0

37

The equation $ax^2 + bx + c = 0$ has two roots, and these are given by

$$x_1 = \frac{-b + \sqrt{b^2 - 4ac}}{2a}, \qquad x_2 = \frac{-b - \sqrt{b^2 - 4ac}}{2a}.$$

The subscripts 1 and 2 serve merely to identify the two roots. They can be omitted, and the above two equations can be summarized by

$$x = \frac{-b \pm \sqrt{b^2 - 4ac}}{2a}.$$

We will not prove these results, though they can be checked by substituting the values for x in the original equation.

Here is a practice problem on finding roots: Which answer correctly gives the roots of $3x - 2x^2 = 1$?

(a) $\frac{1}{4}(3 + \sqrt{17})$; $\frac{1}{4}(3 - \sqrt{17})$
(b) -1; $-\frac{1}{2}$
(c) $\frac{1}{4}$; $-\frac{1}{4}$
(d) 1; $\frac{1}{2}$

Encircle the letter of the correct answer.

[*a* | *b* | *c* | *d*]

If you got the right answer, go to **39**.
If you missed this, go to **38**.

38

Here is the solution to the problem in frame **37**.

The equation $3x - 2x^2 = 1$ can be written in the standard form

$$2x^2 - 3x + 1 = 0.$$

(continued)

Here $a = 2$, $b = -3$, $c = 1$.

$$x = \frac{1}{2a}\left[-b \pm \sqrt{b^2 - 4ac}\right] = \frac{1}{4}\left[-(-3) \pm \sqrt{3^2 - 4 \times 2 \times 1}\right]$$

$$= \frac{1}{4}(3 \pm 1).$$

$$x_1 = \frac{1}{4}(3 + 1) = \frac{1}{4} \times 4 = 1.$$

$$x_2 = \frac{1}{4}(3 - 1) = \frac{1}{4} \times 2 = \frac{1}{2}.$$

Go to **39**.

39

This ends our brief discussion of linear and quadratic functions. Perhaps you would like some more practice on these topics before continuing. If so, try working review problems 1–5 at the back of the book. In Chapter 4 there is a concise summary of the material we have had so far, which you may find useful.

Whenever you are ready, go to **40**.

Trigonometry

40

Trigonometry involves angles, so here is a quick review of the units we use to measure angles. There are two important units: degrees and radians.

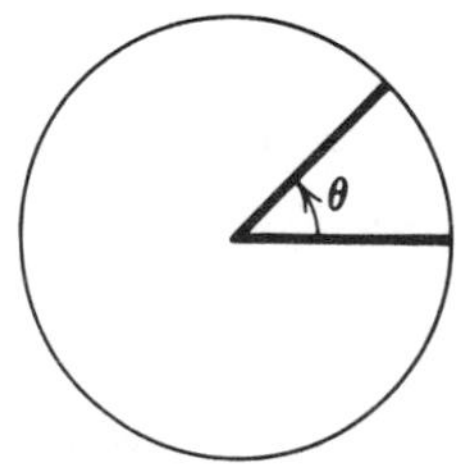

Degrees: Angles are often measured in degrees with 360 degrees (written 360°) corresponding to one complete revolution. [The degree is further subdivided into 60 minutes (60′), and the minute is subdivided

---

Answer: (37) d

into 60 seconds (60"). However, we will not need to use such fine divisions here.] It follows from this that a semicircle contains 180°. Which of the following angles is equal to the angle θ (Greek letter theta) shown in the figure?

[25° | 45° | 90° | 180°]

If right, go to **42**.
Otherwise, go to **41**.

41

To find the angle θ, let's first look at a related example.

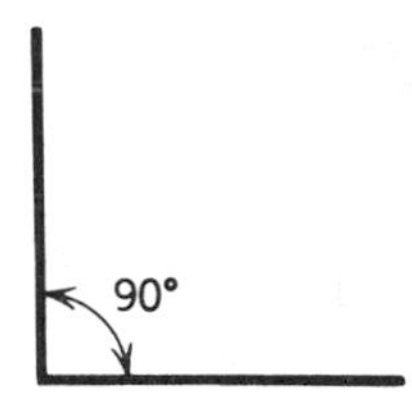

The angle shown is a right angle. Since there are four right angles in a full revolution, it is apparent that the angle equals

$$\frac{360°}{4} = 90°.$$

The angle θ shown in frame **40** is just half as big as the right angle; thus it is 45°.

Here is a circle divided into equal segments by three straight lines. Which angle equals 240°?

[a | b | c]

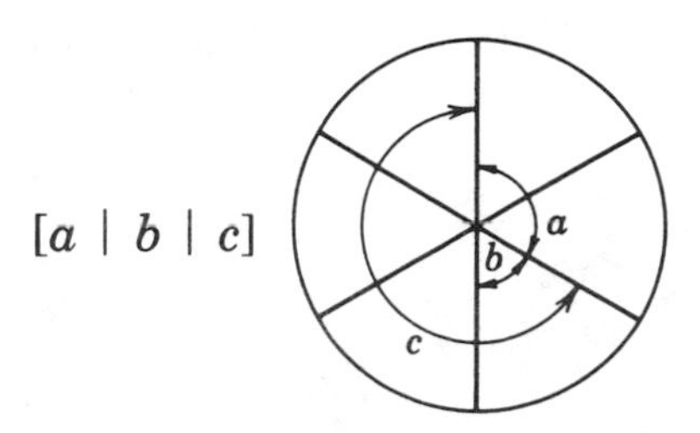

Go to **42**.

42

The second unit of angular measure, and the most useful for calculus, is the *radian*.

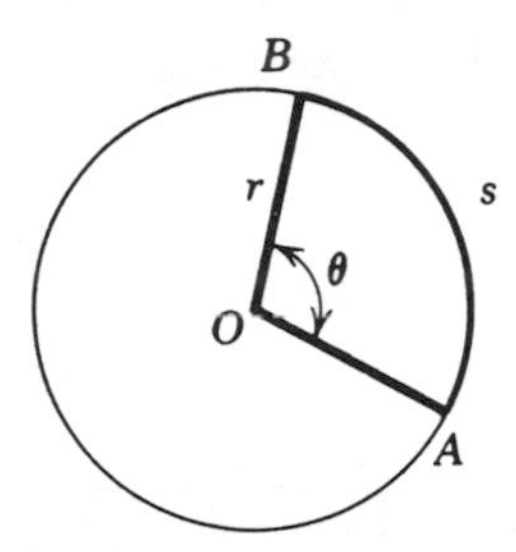

To find the value of an angle in radians, we draw a circle of radius r, about the vertex, O, of the angle so that it intersects the sides of the angle at two points, shown in the figure as A and B. The length of the arc between A and B is designated by s. Then,

$$\theta \text{ (in radians)} = \frac{s}{r} = \frac{\text{length of arc}}{\text{radius}}.$$

To see whether you have caught on, answer this question: There are 360 degrees in a circle; how many radians are there?

[1 | 2 | π | 2π | $360/\pi$]

If right, go to **44**.
Otherwise, go to **43**.

43

The circumference of a circle is πd or $2\pi r$, where d is the diameter and r is the radius.

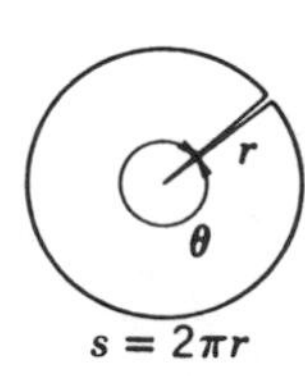

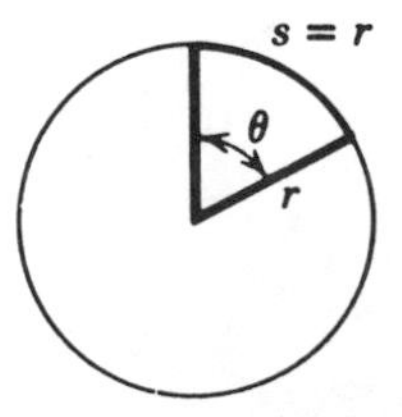

The length of an arc going completely around a circle is the circumference, $2\pi r$, so the angle enclosed is $2\pi r/r = 2\pi$ radians, as shown in the

Answers: (40) 45° **(41)** c

figure on the left. In the figure on the right the angle θ subtends an arc $s = r$. Encircle the answer which gives θ.

[1 rad | ¼ rad | ½ rad | π rad | none of these]

Go to **44**.

44

Because many of the relations we develop later are much simpler when the angles are measured in radians, we will stick to the rule that *all angles will be in radians unless they are marked in degrees.*

Sometimes the word radian is written in full, sometimes it is abbreviated to rad, but usually it is omitted entirely. Thus: $\theta = 0.6$ means 0.6 radian; 27° means 27 degrees; $\pi/3$ rad means $\pi/3$ radians.

Go on to **45**.

45

Since 2π rad = 360°, the rule for converting angles from degrees to radians is

$$1 \text{ rad} = \frac{360°}{2\pi}.$$

Conversely,

$$1° = \frac{2\pi \text{ rad}}{360}.$$

Try the following problems (encircle the correct answer):

60° = [$2\pi/3$ | $\pi/3$ | $\pi/4$ | $\pi/6$] rad

$\pi/4$ = [22½° | 45° | 60° | 90°]

Which angle is closest to 1 rad? (Remember that $\pi = 3.14\ldots$.)

[30° | 45° | 60° | 90°]

If right, go to **47**.
If you made any mistake, go to **46**.

46

Here are the solutions to the problems in frame **45**. From the formulas in frame **45**, one obtains

$$60° = 60 \times \frac{2\pi \text{ rad}}{360} = \frac{2\pi \text{ rad}}{6} = \frac{\pi}{3} \text{ rad.}$$

$$\frac{\pi}{4}\text{rad} = \frac{\pi}{4} \times \frac{360°}{2\pi} = \frac{360°}{8} = 45°.$$

$$1 \text{ rad} = \frac{360°}{2\pi}.$$

Since 2π is just a little greater than 6, 1 rad is slightly less than 360°/6 = 60°. (A closer approximation to the radian is 57°18'.) The figure below shows all the angles in this question.

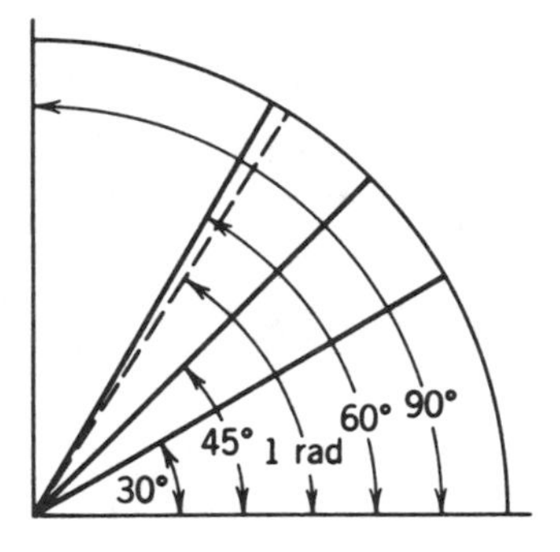

Go to **47**.

47

In the circle shown, CG is perpendicular to AE and

arc AB = arc BC = arc AH,

arc AD = arc DF = arc FA.

Answers: (42) 2π **(43)** 1 rad
(45) $\pi/3$, 45 degrees, 60°

(Arc AB means the length of the arc along the circle between A and B, going the shortest way.)

We will designate angles by three letters. For example, $\angle AOB$ (read as "angle AOB") designates the angle between OA and OB. Try the following:

$$\angle AOD = [60° \mid 90° \mid 120° \mid 150° \mid 180°]$$

$$\angle FOH = [15° \mid 30° \mid 45° \mid 60° \mid 75° \mid 90 \text{ degrees}]$$

$$\angle HOB = [1/4 \mid 1 \mid \pi/2 \mid \pi/4 \mid \pi/8]$$

If you did all these correctly, go to **49**.
If you made any mistakes, go to **48**.

48

Since arc AD = arc DF = arc FA, and since the sum of their angles is 60°, $\angle AOD = 360°/3 = 120°$.

$$\angle FOA = 120°, \qquad \angle GOA = 90°, \qquad \angle GOH = 45°.$$

Thus

$$\angle FOH = \angle FOG + \angle GOH = 30° + 45° = 75°.$$

$$\angle HOB = \angle HOA + \angle AOB = 45° + 45° = 90°.$$

Now try the following:

$$90° = [2\pi \mid \pi/6 \mid \pi/2 \mid \pi/8 \mid 1/4]$$

$$3\pi = [240° \mid 360° \mid 540° \mid 720°]$$

$$\pi/6 = [15° \mid 30° \mid 45° \mid 60° \mid 90° \mid 120°]$$

Go to **49**.

49

Rotations can be counterclockwise or clockwise. By choosing a convention for the sign of an angle, we can indicate which direction is meant. An

(continued)

angle formed by rotating in a counterclockwise direction is positive; an angle formed by moving in a clockwise direction is negative.

Here is a circle of radius r drawn with x- and y-axes, as shown:

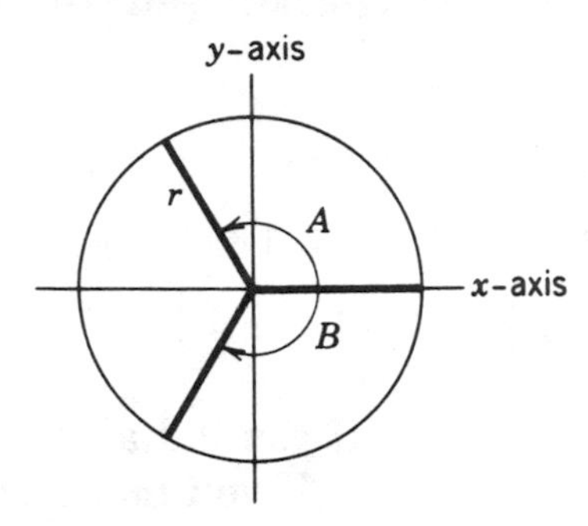

We will choose the positive x-axis as the initial scale and, for the purpose of this section, we will measure angles from the initial to the final or terminal side. As an example, the angle A is positive and B is negative, as shown in the figure.

Go to **50**.

50

Our next task is to review the trigonometric functions. One use of these functions is to relate the sides of triangles, particularly right triangles, to their angles.

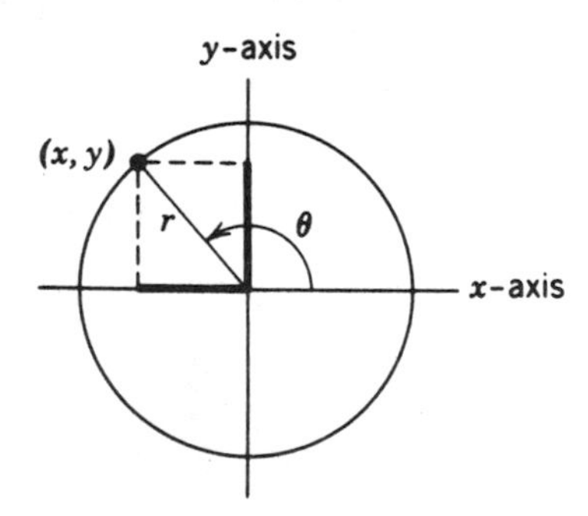

We will get to this application shortly. However, the trigonometric functions can be defined in a more general and more useful way.

Do you know the general definitions of the trigonometric functions of

---

Answers: (47) 120°, 75°, $\pi/2$
(48) $\pi/2$, 540°, 30°

angle θ? If you do, test yourself with the quiz below. If you don't, go right on to frame **51**.

The trigonometric functions of θ can be expressed in terms of the coordinates x and y and the radius of the circle, $r = \sqrt{x^2 + y^2}$. These are shown in the figure. Try to fill in the blanks (the answers are in frame **51**):

$\sin\theta =$ ______ $\quad \cot\theta =$ ______

$\cos\theta =$ ______ $\quad \sec\theta =$ ______

$\tan\theta =$ ______ $\quad \csc\theta =$ ______

Go to frame **51** to check your answers.

51

Here are the definitions of the trigonometric functions:

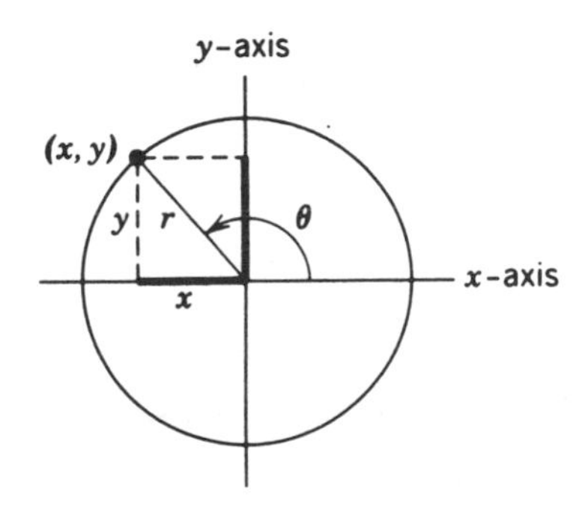

sine: $\sin\theta = \dfrac{y}{r}$, $\quad$ cotangent: $\cot\theta = \dfrac{1}{\tan\theta} = \dfrac{x}{y}$,

cosine: $\cos\theta = \dfrac{x}{r}$, $\quad$ secant: $\sec\theta = \dfrac{1}{\cos\theta} = \dfrac{r}{x}$,

tangent: $\tan\theta = \dfrac{y}{x}$, $\quad$ cosecant: $\csc\theta = \dfrac{1}{\sin\theta} = \dfrac{r}{y}$.

Note that the last three are merely reciprocals of the first three. For the angle shown in the figure, x is negative and y is positive ($r = \sqrt{x^2 + y^2}$ and is always positive) so that $\cos\theta$, $\tan\theta$, $\cot\theta$, and $\sec\theta$ are negative.

After you have studied these, go to **52**.

52

Below is a circle with a radius of 5. The point shown is (−3, −4). On the basis of the definition in the last frame, you should be able to answer the following:

$\sin\,\theta = [3/5 \mid 5/3 \mid 3/4 \mid -4/5 \mid -3/5 \mid 4/3]$

$\cos\,\theta = [3/5 \mid 5/3 \mid 3/4 \mid -4/5 \mid -3/5 \mid 4/3]$

$\tan\,\theta = [3/5 \mid 5/3 \mid 3/4 \mid -4/5 \mid -3/5 \mid 4/3]$

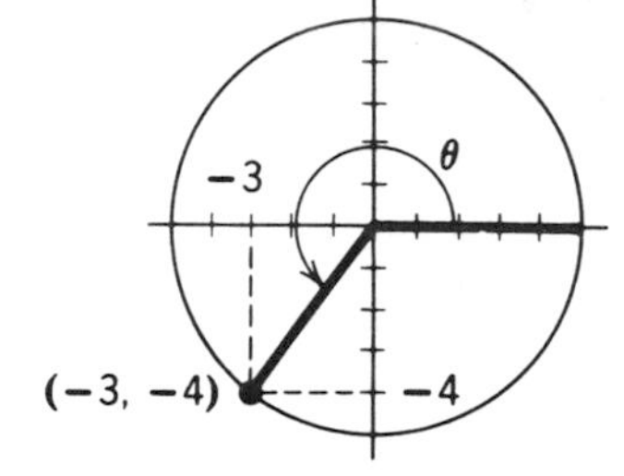

If all right, go to **55**.
Otherwise, go to **53**.

53

Perhaps you had difficulty because you did not realize that x and y have different signs in different quadrants (quarters of the circle) while r, a radius, is always positive. Try this problem.

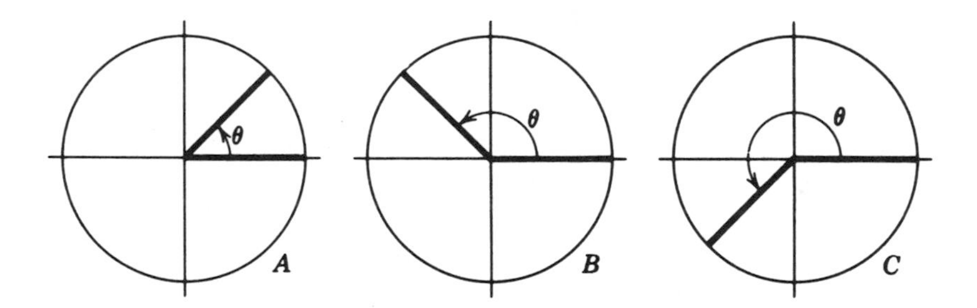

Indicate whether the function required is positive or negative, for each of the figures, by checking the correct box.

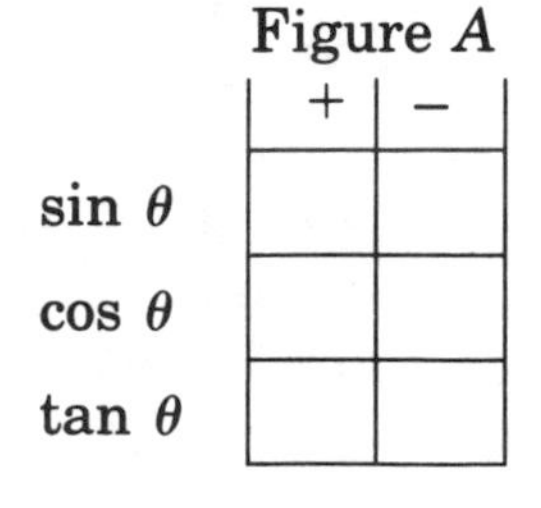

	Figure *A* +	Figure *A* −
sin θ		
cos θ		
tan θ		

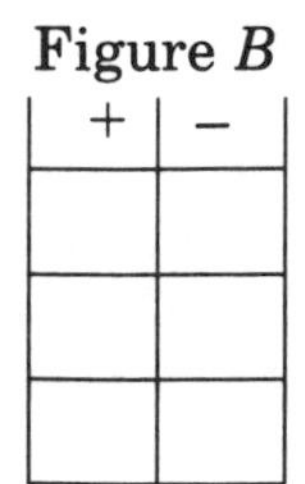

	Figure *B* +	Figure *B* −
sin θ		
cos θ		
tan θ		

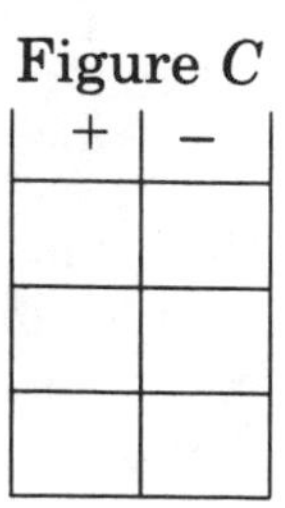

	Figure *C* +	Figure *C* −
sin θ		
cos θ		
tan θ		

See frame **54** for the correct answers.

54

Here are the answers to the questions in frame **53**.

Figure A

	+	–
$\sin\theta$	√	
$\cos\theta$	√	
$\tan\theta$	√	

Figure B

	+	–
$\sin\theta$	√	
$\cos\theta$		√
$\tan\theta$		√

Figure C

	+	–
$\sin\theta$		√
$\cos\theta$		√
$\tan\theta$	√	

Go to **55**.

55

In the figure both θ and $-\theta$ are shown. The trigonometric functions for these two angles are simply related. Can you do these problems? Encircle the correct sign.

$\sin(-\theta) = [+ \mid -] \sin\theta$

$\cos(-\theta) = [+ \mid -] \cos\theta$

$\tan(-\theta) = [+ \mid -] \tan\theta$

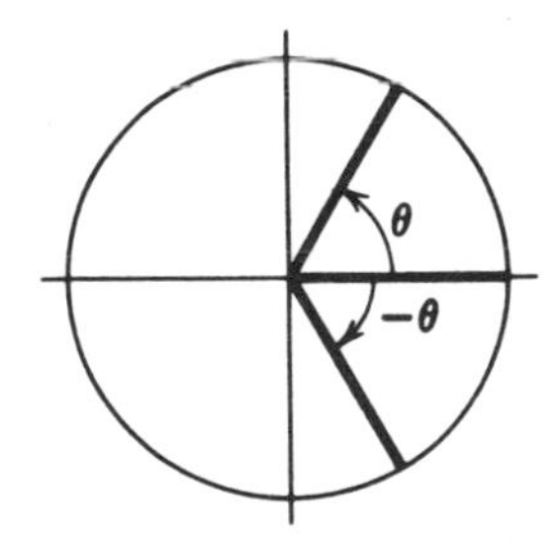

Go to **56**.

56

There are many relationships among the trigonometric functions. For instance, using $x^2 + y^2 = r^2$, we have

$$\sin^2\theta = \frac{y^2}{r^2} = \frac{r^2 - x^2}{r^2} = 1 - \left(\frac{x}{r}\right)^2 = 1 - \cos^2\theta.$$

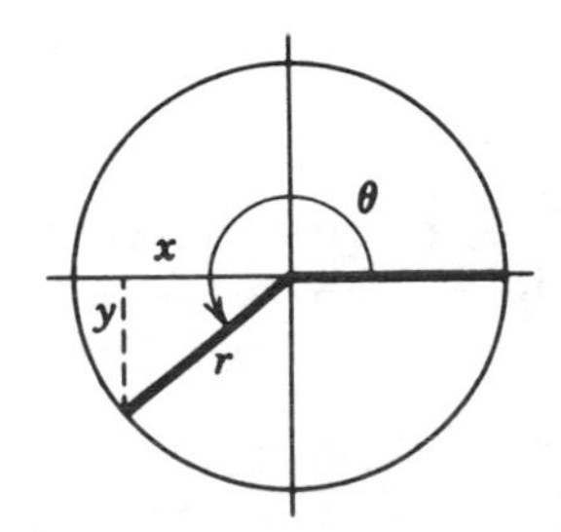

(continued)

Try these:

1. $\sin^2\theta + \cos^2\theta = [\sec^2\theta \mid 1 \mid \tan^2\theta \mid \cot^2\theta]$
2. $1 + \tan^2\theta = [1 \mid \tan^2\theta \mid \cot^2\theta \mid \sec^2\theta]$
3. $\sin^2\theta - \cos^2\theta = [1 - 2\cos^2\theta \mid 1 - 2\sin^2\theta \mid \cot^2\theta \mid 1]$

If any mistakes, go to **57**.
Otherwise, go to **58**.

57

Here are the solutions to the problems in frame **56**.

1. $\sin^2\theta + \cos^2\theta = \frac{y^2}{r^2} + \frac{x^2}{r^2} = \frac{x^2 + y^2}{r^2} = \frac{r^2}{r^2} = 1.$

This is an important identity which is worth remembering. The other solutions are

2. $1 + \tan^2\theta = 1 + \frac{\sin^2\theta}{\cos^2\theta} = \frac{\cos^2\theta + \sin^2\theta}{\cos^2\theta} = \frac{1}{\cos^2\theta} = \sec^2\theta.$
3. $\sin^2\theta - \cos^2\theta = 1 - \cos^2\theta - \cos^2\theta = 1 - 2\cos^2\theta.$

Go to **58**.

58

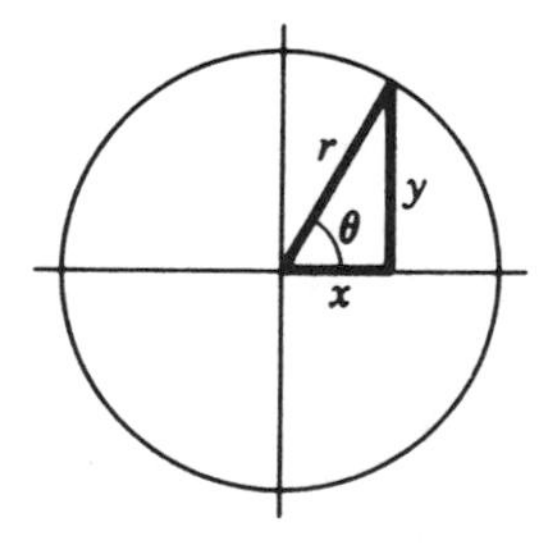

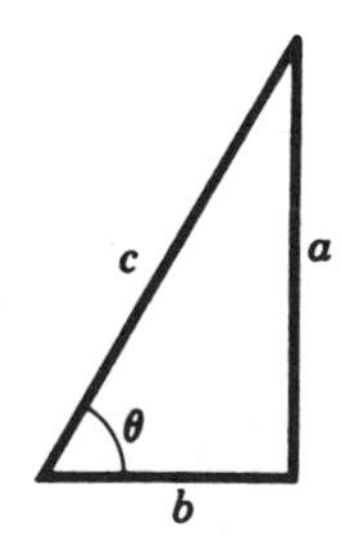

The trigonometric functions are particularly useful when applied to right triangles (triangles with one 90° or right angle). In this case θ is

---

Answers: (52) $-4/5, -3/5, 4/3$
(55) $-, +, -$

always acute (less than 90° or $\pi/2$). You can then write the trigonometric functions in terms of the sides a, b of the right triangle shown, and its hypotenuse c. Fill in the blanks.

$\sin\theta =$ ________ $\quad \cot\theta =$ ________

$\cos\theta =$ ________ $\quad \sec\theta =$ ________

$\tan\theta =$ ________ $\quad \csc\theta =$ ________

Check your answer in **59**.

59

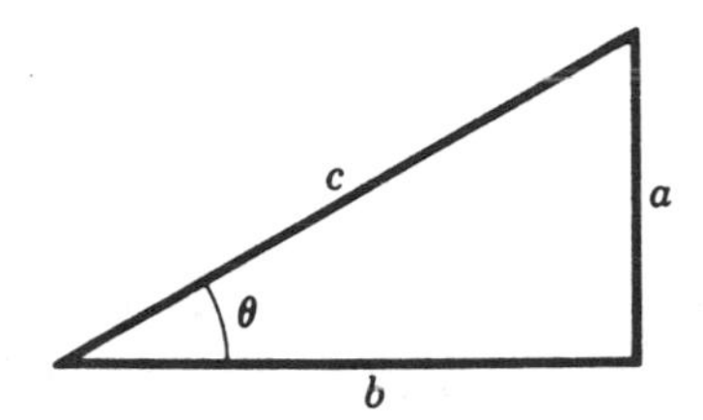

The answers are:

$$\sin\theta = \frac{a}{c} = \frac{\text{opposite side}}{\text{hypotenuse}}, \qquad \cot\theta = \frac{b}{a} = \frac{\text{adjacent side}}{\text{opposite side}},$$

$$\cos\theta = \frac{b}{c} = \frac{\text{adjacent side}}{\text{hypotenuse}}, \qquad \sec\theta = \frac{c}{b} = \frac{\text{hypotenuse}}{\text{adjacent side}},$$

$$\tan\theta = \frac{a}{b} = \frac{\text{opposite side}}{\text{adjacent side}}, \qquad \csc\theta = \frac{c}{a} = \frac{\text{hypotenuse}}{\text{opposite side}}.$$

These results follow from the definitions in frame **51**, providing we let a, b, and c correspond to y, x, and r, respectively. (Remember that here θ is less than 90°.) If you are not familiar with the terms opposite side, adjacent side, and hypotenuse, they should be evident from the figure.

Go to **60**.

60

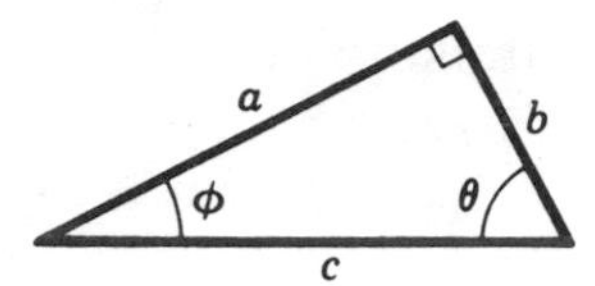

The following problems refer to the figure shown. (ϕ is the Greek letter "phi." The symbol between sides a and b indicates a right angle.)

$$\cos\theta = [b/c \mid a/c \mid c/a \mid c/b \mid b/a \mid a/b]$$

$$\tan\theta = [b/c \mid a/c \mid c/a \mid c/b \mid b/a \mid a/b]$$

If all right, go to **62**.
Otherwise, go to **61**.

61

You may have become confused because the triangle was drawn in a new position. Review the definitions in **51**, and then do the problems below:

$$\sin\theta = [l/n \mid n/l \mid m/n \mid m/l \mid n/m \mid l/m]$$

$$\tan\phi = [l/n \mid n/l \mid m/n \mid m/l \mid n/m \mid l/m]$$

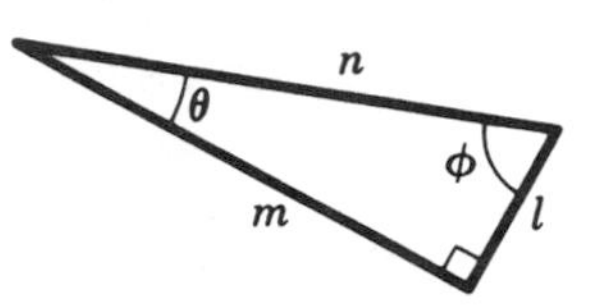

If you missed either of these, you will have to put in more work learning and memorizing the definitions.

Meanwhile go to **62**.

62

It is helpful to be familiar with the trigonometric functions of 30°, 45°, and 60°. The triangles for these angles are particularly simple.

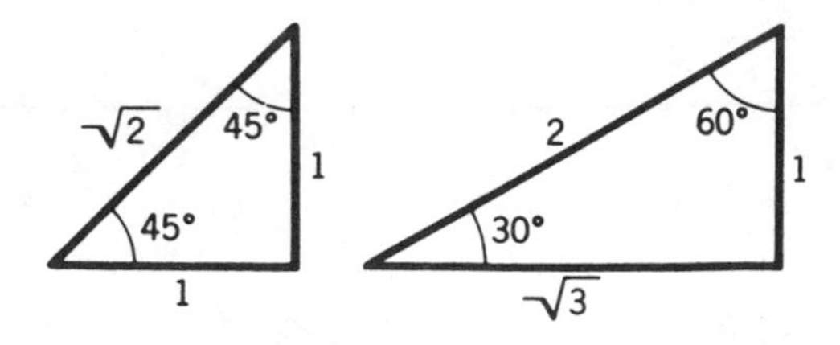

Answers: (56) 1, $\sec^2\theta$, $1 - 2\cos^2\theta$

Try these problems:

$\cos 45° = [1/2 \mid 1/\sqrt{2} \mid 2\sqrt{2} \mid 2]$

$\sin 30° = [3 \mid \sqrt{3/2} \mid 2/3 \mid 1/2]$

$\sin 45° = [1/2 \mid 1/\sqrt{2} \mid \sqrt{2/2} \mid 2]$

$\tan 30° = [1 \mid \sqrt{3} \mid 1/\sqrt{3} \mid 2]$

Make sure you understand these problems. Then go to **63**.

63

Many calculators provide values of trigonometric functions. With such a calculator, it is quite simple to plot enough points to make a good graph of the function. If you have such a calculator, plot sin θ for values between 0° and 360° on the coordinate axes below, and then compare your result with frame **64**. If you do not have a suitable calculator, go directly to **64** and check that sin θ has the correct values for the angles you know.

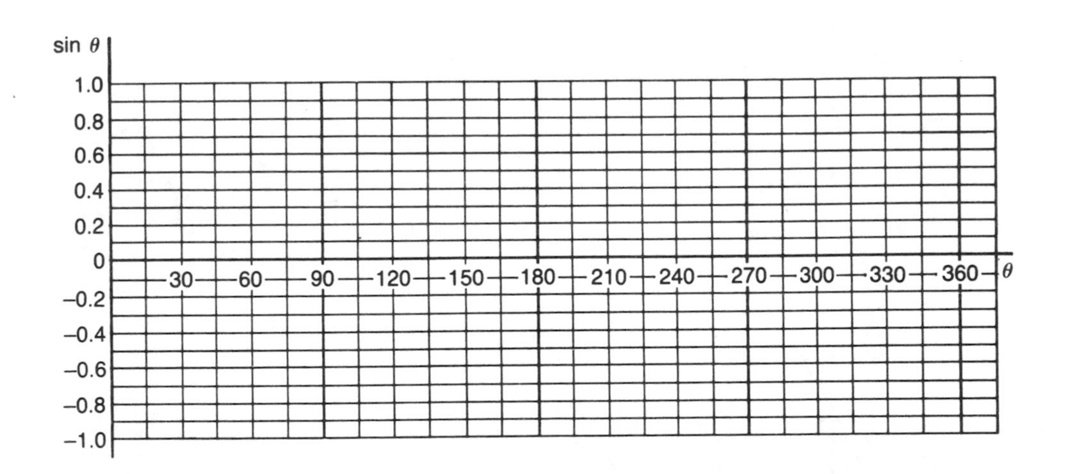

Go to **64**.

64

Here is the graph of the sine function.

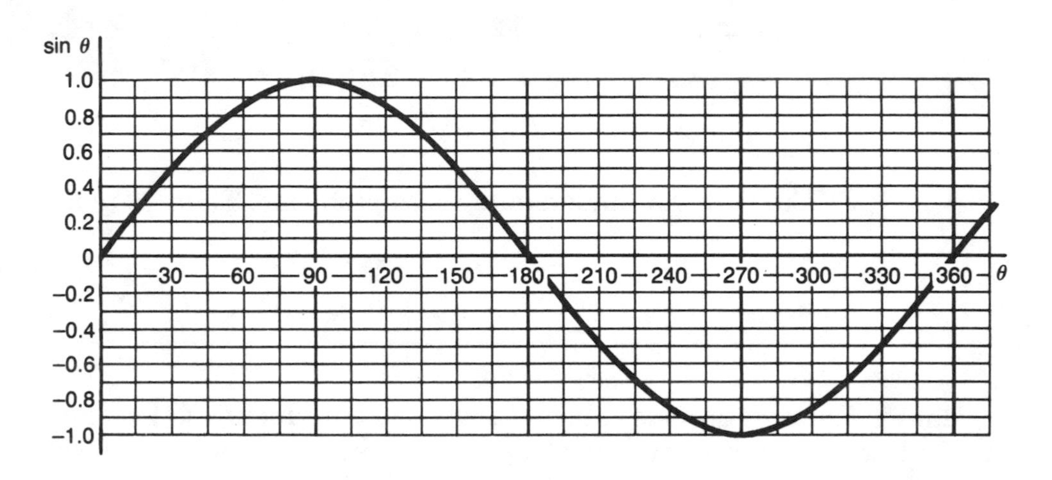

Go to **65**.

65

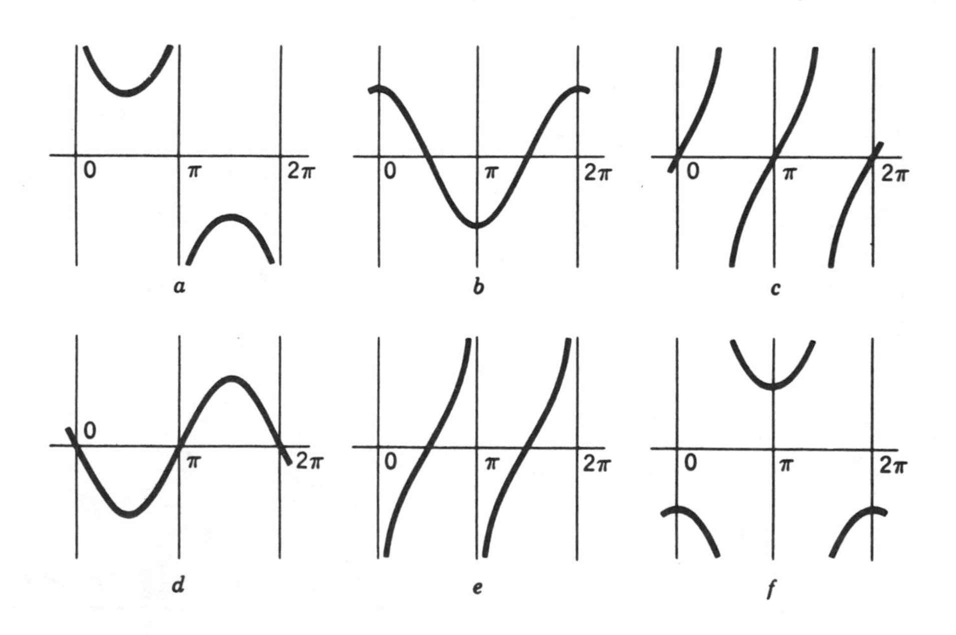

Answers: (60) $b/c, a/b$

(61) l/n, m/l **(62)** $1/\sqrt{2}$, $1/2$, $1/\sqrt{2}$, $1/\sqrt{3}$

Try to decide which graph represents each function.

cos θ: [a | b | c | d | e | f | none of these]

tan θ: [a | b | c | d | e | f | none of these]

sin($-\theta$): [a | b | c | d | e | f | none of these]

tan($-\theta$): [a | b | c | d | e | f | none of these]

If you got these all right, go to **67**.
Otherwise go to **66**.

66

Knowing the values of the trigonometric functions at a few important points will help you identify them. Try these (∞ is the symbol for infinity):

sin 0° = [0 | 1 | −1 | −∞ | +∞]

cos 90° = [0 | 1 | −1 | −∞ | +∞]

tan 45° = [0 | 1 | −1 | −∞ | +∞]

sin 30° = [1 | 1/2 | $\sqrt{3}$ | $\sqrt{3}/2$]

cos 60° = [1 | 1/2 | $\sqrt{3}$ | $\sqrt{3}/2$]

Go to **67**.

67

Because the angle $\theta + 2\pi$ is equivalent to θ as far as the trigonometric functions are concerned, we can add 2π to any angle without changing the value of the trigonometric functions. Thus, the sine and cosine functions repeat their values whenever θ increases by 2π; we say that the functions are *periodic* in θ with a period of 2π or with a period of 360°.

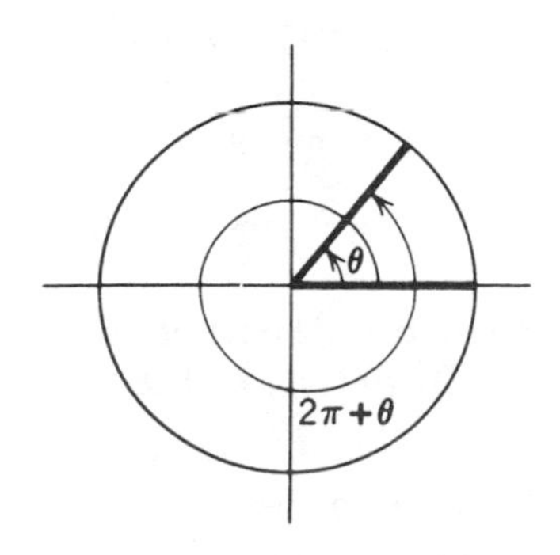

(continued)

Using this property, you can extend the graph of sin θ in frame **64** to the following. (For variety, the angle here is in radians.)

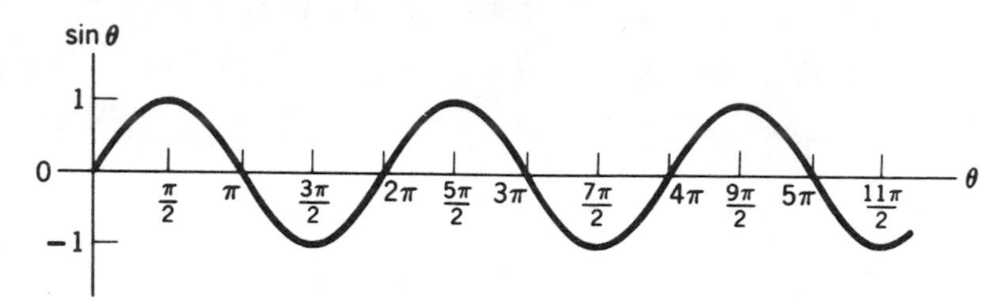

Go to **68**.

68

It is helpful to know the sine and cosine of the sum and the difference of two angles.

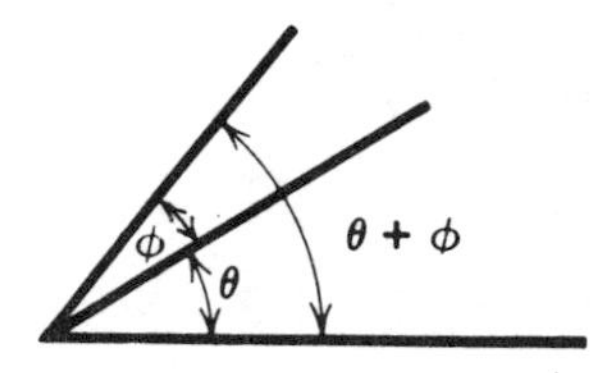

Do you remember the formulas from previous studies of trigonometry? If not, go to **69**. If you do, try the quiz below.

$\sin(\theta + \phi) =$ ________________________.

$\cos(\theta + \phi) =$ ________________________.

Go to **69** to see the correct answer.

69

Here are the formulas. They are derived in Appendix A1.

$$\sin(\theta + \phi) = \sin\theta\cos\phi + \cos\theta\sin\phi,$$

$$\cos(\theta + \phi) = \cos\theta\cos\phi - \sin\theta\sin\phi.$$

[Note that $\tan(\theta + \phi)$ and $\cot(\theta + \phi)$ can be obtained from these formulas and the relation $\tan\theta = (\sin\theta)/(\cos\theta)$.]

Answers: (65) *b*, *c*, *d*, none of these;
(66) 0, 0, 1, ½, ½

By using what you have already learned, circle the correct sign in each of the following:

(a) $\sin(\theta - \phi) = [+ \mid -] \sin\theta \cos\phi \, [+ \mid -] \cos\theta \sin\phi$

(b) $\cos(\theta - \phi) = [+ \mid -] \cos\theta \cos\phi \, [+ \mid -] \sin\theta \sin\phi$

If right, go to **71**.
If wrong, go to **70**.

70

If you made a mistake in problem **68**, you should recall from frame **55** that

$$\sin(-\phi) = -\sin\phi,$$

$$\cos(-\phi) = +\cos\phi.$$

Then

$$\begin{aligned}\sin(\theta - \phi) &= \sin\theta\cos(-\phi) + \cos\theta\sin(-\phi)\\ &= \sin\theta\cos\phi - \cos\theta\sin\phi,\end{aligned}$$

$$\begin{aligned}\cos(\theta - \phi) &= \cos\theta\cos(-\phi) - \sin\theta\sin(-\phi)\\ &= \cos\theta\cos\phi + \sin\theta\sin\phi.\end{aligned}$$

Go to **71**.

71

By using the expressions for $\sin(\theta + \phi)$ and $\cos(\theta + \phi)$, one can obtain the formulas for $\sin 2\theta$ and $\cos 2\theta$. Simply let $\theta = \phi$. Fill in the blanks.

$$\sin 2\theta = \underline{\hspace{6cm}}.$$

$$\cos 2\theta = \underline{\hspace{6cm}}.$$

See **72** for the correct answers.

72

$$\sin 2\theta = 2 \sin \theta \cos \theta,$$

$$\cos 2\theta = \cos^2 \theta - \sin^2 \theta$$
$$= 1 - 2 \sin^2 \theta$$
$$= 2 \cos^2 \theta - 1.$$

[Note, by convention, $(\sin \theta)^2$ is usually written $\sin^2 \theta$, and $(\cos \theta)^2$ is usually written $\cos^2 \theta$.]

Go to **73**.

73

It is often useful and convenient to use the *inverse trigonometric function,* which is the value of the angle for which the trigonometric function has a specified value. The inverse sine of x is denoted by $\sin^{-1} x$. (Warning: This notation is standard, but it can be confusing. $\sin^{-1} x$ always represents the inverse sine of x, *not* $1/\sin x$. The latter would be written $(\sin x)^{-1}$. An older notation for $\sin^{-1} x$ is $\arcsin x$.)

For example, since the sine of 30° is ½, $\sin^{-1} \frac{1}{2} = 30°$. Note, however, that the sine of 150° is also ½. Furthermore, the trigonometric functions are periodic: there is an endless sequence of angles (all differing by 360°) having the same value for the sine, cosine, etc.

Because the definition of function (frame **6**) specifies the assignment of one and only one value of $f(x)$ for each value of x, the range of the inverse trigonometric function must be suitably restricted.

The inverse functions are defined by

$y = \sin^{-1} x$	Domain: $-1 < x < +1$	Range: $-\frac{\pi}{2} < y < +\frac{\pi}{2}$
$y = \cos^{-1} x$	Domain: $-1 < x < +1$	Range: $0 < y < \pi$
$y = \tan^{-1} x$	Domain: $-\infty < x < +\infty$	Range: $-\frac{\pi}{2} < y < +\frac{\pi}{2}$

Go to **74**.

Answers: (69a) +, −; **(69b)** +, +

74

Try these problems:

(a) $\sin^{-1}(1/\sqrt{2})$ = [30° | 60° | $\pi/4$ | $\pi/2$]

(b) $\tan^{-1} 1$ = [$\pi/6$ | $\pi/4$ | $\pi/3$ | π]

(c) $\cos^{-1}(1/2)$ = [$\pi/6$ | $\pi/4$ | $\pi/3$ | π]

If you have a calculator with inverse trigonometric functions, try the following:

(d) $\sin^{-1} 0.8$ = [46.9 | 28.2 | 53.1 | 67.2] degrees

(e) $\tan^{-1} 12$ = [0.82 | 1.49 | 1.62 | 1.83] radians

(f) $\cos^{-1} 0.05$ = [4.3 | 12.6 | 77.2 | 87.1] degrees

Check your answers, and then go on to the next section, which is the last one in our reviews.

Go to **75**.

Exponentials and Logarithms

75

Are you already familiar with exponentials? If not, go to **76**. If you are, try this short quiz.

a^5 = [5^a | $5 \log a$ | $a \log 5$ | none of these]

a^{b+c} = [$a^b \times a^c$ | $a^b + a^c$ | ca^b | $(b + c) \log a$]

a^f/a^g = [$(f - g) \log a$ | $a^{f/g}$ | a^{f-g} | none of these]

a^0 = [0 | 1 | a | none of these]

$(a^b)^c$ = [$a^b \times a^c$ | a^{b+c} | a^{bc} | none of these]

If any mistakes, go to **76**.
Otherwise, go to **77**.

76

By definition a^m, where m is a positive integer, is the product of m factors of a. Hence,

$$2^3 = 2 \times 2 \times 2 = 8 \qquad \text{and} \qquad 10^2 = 10 \times 10 = 100.$$

Furthermore, by definition $a^{-m} = 1/a^m$. It is easy to see, then, that

$$a^m \times a^n = a^{m+n},$$
$$\frac{a^m}{a^n} = a^{m-n},$$
$$a^0 = \frac{a^m}{a^m} = 1 \qquad (m \text{ can be any integer}),$$
$$(a^m)^n = a^{mn},$$
$$(ab)^m = a^m b^m.$$

Note that a^{m+n} is evaluated as $a^{(m+n)}$; the expression in the exponential is always evaluated before any other operation is carried out.

If you have not yet tried the quiz in frame **75**, go to **75**. Otherwise,

Go to **77**.

77

Here are a few problems:

$$3^2 = [6 \mid 8 \mid 9 \mid \text{none of these}]$$
$$1^3 = [1 \mid 3 \mid \frac{1}{3} \mid \text{none of these}]$$
$$2^{-3} = [-6 \mid \frac{1}{8} \mid -9 \mid \text{none of these}]$$
$$\frac{4^3}{4^5} = [4^8 \mid 4^{-8} \mid 16^{-1} \mid \text{none of these}]$$

If you did these all correctly, go to **79**.
If you made any mistakes, go to **78**.

Answers: **(74)** (a) $\pi/4$, (b) $\pi/4$, (c) $\pi/3$, (d) 53.1°, (e) 1.49, (f) 87.1°
(75) None of these, $a^b \times a^c$, a^{f-g}, 1, a^{bc}

78

Below are the solutions to problem **77**. Refer back to the rules in **76** if you have trouble understanding the solution.

$$3^2 = 3 \times 3 = 9,$$
$$1^3 = 1 \times 1 \times 1 = 1 \qquad (1^m = 1 \text{ for any } m),$$
$$2^{-3} = \frac{1}{2^3} = \frac{1}{8},$$
$$\frac{4^3}{4^5} = 4^{3-5} = 4^{-2} = \frac{1}{16} = 16^{-1}.$$

Now try these:

$$(3^{-3})^3 = [1 \mid 3^{-9} \mid 3^{-27} \mid \text{none of these}]$$
$$\frac{5^2}{3^2} = [\left(\frac{5}{3}\right)^2 \mid \left(\frac{5}{3}\right)^{-1} \mid 5^{-6} \mid \text{none of these}]$$
$$4^3 = [12 \mid 16 \mid 2^6 \mid \text{none of these}]$$

Check your answers and try to track down any mistakes.

Then go to **79**.

79

Here are a few more problems.

$$10^0 = [0 \mid 1 \mid 10]$$
$$10^{-1} = [-1 \mid 1 \mid 0.1]$$
$$0.00003 = [\frac{1}{3} \times 10^{-3} \mid 10^{-3} \mid 3 \times 10^{-5}]$$
$$0.4 \times 10^{-4} = [4 \times 10^{-5} \mid 4 \times 10^{-3} \mid 2.5 \times 10^{-5}]$$
$$\frac{3 \times 10^{-7}}{6 \times 10^{-3}} = [\frac{1}{2} \times 10^{10} \mid 5 \times 10^4 \mid 0.5 \times 10^{-4}]$$

If these were all correct, go to **81**.
If you made any mistakes, go to **80**.

80

Here are the solutions to the problems in **79**:

$$10^0 = \frac{1}{10} = 0.1,$$

$$0.00003 = 0.00001 \times 3 = 3 \times 10^{-5},$$

$$0.4 \times 10^{-4} = (4 \times 10^{-1}) \times 10^{-4} = 4 \times 10^{-5},$$

$$\frac{3 \times 10^{-7}}{6 \times 10^{-3}} = \frac{3}{6} \times \frac{10^{-7}}{10^{-3}} = \frac{1}{2} \times 10^{-7+3} = 0.5 \times 10^{-4}.$$

Go to **81**.

81

Let's briefly review fractional exponents. If $b^n = a$, then b is called the nth root of a and is written $b = a^{1/n}$. Hence $16^{1/4}$ = (fourth root of 16) = 2. That is, $2^4 = 16$.

If $y = a^{m/n}$, where m and n are integers, then $y = [a^{1/n}]^m$. For instance.

$$8^{2/3} = (8^{1/3})^2 = 2^2 = 4.$$

Try these:

$27^{-2/3}$ = [1/18 | 1/81 | 1/9 | −18 | none of these]

$16^{3/4}$ = [12 | 8 | 6 | 64]

If right, go to **84**.
If wrong, go to **82**.

82

$$27^{-2/3} = (27^{1/3})^{-2} = 3^{-2} = 1/9,$$

$$16^{3/4} = (16^{1/4})^3 = 2^3 = 8.$$

Answers: (77) 9, 1, 1/8, 16^{-1} **(78)** 3^{-9}, $(5/3)^2$, 2^6
(79) 1, 0.1, 3×10^{-5}, 4×10^{-5}, 0.5×10^{-4}

Do these problems:

$$25^{3/2} = [125 \mid 5 \mid 15 \mid \text{none of these}]$$
$$(0.00001)^{-3/5} = [0.001 \mid 1000 \mid 10^{-15} \mid 10^{-25}]$$

If your answers were correct, go to **84**.
Otherwise, go to **83**.

83

Here are the solutions to the problems in **82**.

$$25^{3/2} = (25^{1/2})^3 = 5^3 = 125,$$
$$(0.00001)^{-3/5} = (10^{-5})^{-3/5} = 10^{15/5} = 10^3 = 1000.$$

Here are a few more problems. Encircle the correct answers.

$$(\tfrac{27}{64} \times 10^{-6})^{1/3} = [\tfrac{3}{400} \mid \tfrac{3}{16} \times 10^{-2} \mid \tfrac{9}{64} \times 10^{-4}],$$
$$(49 \times 10^{-4})^{1/4} = [\sqrt{7}/10 \mid (10 \times 7)^{-2} \mid \sqrt{7}/1000].$$

Go to **84** after checking your answers.

84

Although our original definition of a^m only applied to integral values of m, we have also defined $(a^m)^{1/n} = a^{m/n}$, where both m and n are integers. Thus we have a meaning for a^p, where p is either an integer or a fraction (ratio of integers).

As yet we do not know how to evaluate a^p if p is an irrational number, such as π or $\sqrt{2}$. However, we can approximate an irrational number as closely as we desire by fraction. For instance, π is approximately 31,416/10,000. This is in the form m/n, where m and n are integers, and we know how to evaluate it. Therefore, $y = a^x$, where x is any real number, is a meaningful expression in the sense that we can evaluate it as accurately as we please. (A more rigorous treatment of irrational exponents can be based on the properties of suitably defined logarithms.)

Try the following problem.

$$\frac{a^{\pi}a^{x}}{a^3} = [a^{\pi x/3} \mid a^{\pi+x-3} \mid a^{3\pi x} \mid a^{(\pi+x)/3}]$$

If right, go to **86**.
If wrong, go to **85**.

85

The rules given in frame **76** apply here as if all exponents were integers. Hence

$$\frac{a^{\pi}a^{x}}{a^{3}} = a^{\pi+x-3}.$$

Here is another problem:

$$\pi^2 \times 2^{\pi} = [1 \mid (2\pi)^{2\pi} \mid 2\pi^{2+\pi} \mid \text{none of these}]$$

If right, go to **87**.
If wrong, go to **86**.

86

$\pi^2 \times 2^{\pi}$ is the product of two different numbers to two different exponents. None of our rules apply to this and, in fact, there is no way to simplify this expression.

Now go to **87**.

87

If you do not clearly remember logarithms, go to **88**. If you do, try the following test.

Let x be any positive number, and let $\log x$ represent the log of x to the base 10. Then:

$$10^{\log x} = \underline{\qquad\qquad}.$$

Go to **88** for the correct answer.

Answers: **(81)** 1/9, 8 **(82)** 125, 1000
(83) 3/400, $\sqrt{7}/10$ **(84)** $a^{\pi+x-3}$

88

The answer to **87** is x; in fact we will take the logarithm of x to the base 10 to be defined by

$$\boxed{10^{\log x} = x.}$$

That is, the logarithm of a number x is the power to which 10 must be raised to produce the number x itself. This definition only applies for $x > 0$. Here are two examples:

$$100 = 10^2, \qquad \text{so } \log 100 = 2;$$
$$0.001 = 10^{-3}, \qquad \text{so } \log 0.001 = -3.$$

Now try these problems:

$$\log 1{,}000{,}000 = [1{,}000{,}000 \mid 6 \mid 60 \mid 600]$$
$$\log 1 = [0 \mid 1 \mid 10 \mid 100]$$

If right, go to **90**.
If wrong, go to **89**.

89

$$\log 1{,}000{,}000 = \log 10^6 = 6 \qquad (\text{check, } 10^6 = 1{,}000{,}000),$$
$$\log 1 = \log 10^0 = 0 \qquad (\text{check, } 10^0 = 1).$$

Try the following problems:

$$\log(10^4/10^{-3}) = [10^7 \mid 1 \mid 10 \mid 7 \mid 70]$$
$$\log 10^n = [10n \mid n \mid 10^n \mid 10/n]$$
$$\log 10^{-n} = [-10n \mid -n \mid -10^n \mid -10/n]$$

If you had trouble with these, review the material in this section. Make sure you understand these problems.

Then go to **90**.

90

Here are three important relations for manipulating logarithms. a and b are any positive numbers:

$$\log ab = \log a + \log b,$$

$$\log(a/b) = \log a - \log b,$$

$$\log a^n = n \times \log a.$$

If you are familiar with these rules, go to **92**. If you want to see how they are derived,

Go to **91**.

91

We can derive the required rules as follows. From the definition of $\log x$, $a = 10^{\log a}$ and $b = 10^{\log b}$. Consequently, from the properties of exponentials,

$$ab = 10^{\log a} \times 10^{\log b} = 10^{\log a + \log b}.$$

Taking the log of both sides, and again using $\log 10^x = x$, gives

$$\log ab = \log 10^{\log a + \log b} = \log a + \log b.$$

Similarly,

$$a/b = 10^{\log a} 10^{-\log b} = 10^{\log a - \log b}.$$

$$\log(a/b) = \log a - \log b$$

Likewise,

$$a^n = (10^{\log a})^n = 10^{n \log a},$$

so that

$$\log a^n = n \times \log a.$$

Go to **92**.

Answer: (**85**) None of these (**88**) 6, 0, (**89**) 7, n, $-n$

92

Try these problems:

$$\text{If } \log n = -3, \qquad n = [1/3 \mid 1/300 \mid 1/1000]$$

$$10^{\log 100} = [10^{10} \mid 20 \mid 100 \mid \text{none of these}]$$

$$\frac{\log 1000}{\log 100} = [\frac{3}{2} \mid 1 \mid -1 \mid 10]$$

If right, go to **94**.
If wrong, go to **93**.

93

$$10^{\log n} = n, \text{ so if } \log n = -3, n = 10^{-3} = 1/1000.$$

For the same reason,

$$10^{\log 100} = 100.$$

$$\frac{\log 1000}{\log 100} = \frac{\log 10^3}{\log 10^2} = \frac{3}{2}.$$

Try these problems:

$$\tfrac{1}{2} \log 16 = [2 \mid 4 \mid 8 \mid \log 2 \mid \log 4]$$

$$\log(\log 10) = [10 \mid 1 \mid 0 \mid -1 \mid -10]$$

Go to **94**.

94

In this section we have discussed only logarithms to the base 10. However, any positive number except 1 can be used as a base. Bases other than 10 are usually indicated by a subscript. For instance, the logarithm of 8 to the base 2 is written $\log_2 8$. This has the value of 3 since $2^3 = 8$. If our base is denoted by r, then the defining equation for $\log_r x$ is

$$\boxed{r^{\log_r x} = x.}$$

(continued)

All the relations explained in frame **91** are true for logarithms to any base (provided, of course, that the same base is used for all the logarithms in each equation).

We shall later discuss *natural logarithms,* for which the base is the number $e = 2.71828 \ldots$. Natural logarithms are usually designated by the symbol $\ln x = \log_e x$. Many calculators give both $\log x$ [i.e., $\log_{10} x$] and $\ln x$.

Go to **95**.

95

From the definition of logarithm in the last frame we can obtain the rule for changing logarithms from one base to another, for instance from base 10 to the base e.

Take $\log_{10}$ of both sides of the defining equation $e^{\ln x} = x$,

$$\log(e^{\ln x}) = \log x.$$

Because $\log x^n = n \log x$ (frame **91**), this gives

$$\ln x \log e = \log x$$

or

$$\ln x = \frac{\log x}{\log e}.$$

The numerical value of $\log e$ is $2.303 \ldots$, so

$$\ln x = \frac{\log x}{2.303 \ldots}.$$

If you have a calculator which evaluates both $\ln x$ and $\log x$, check this relation for a few values of x.

Go to **96**.

Answers: (92) $1/1000$, 100, $3/2$ **(93)** log 4, 0

96

This concludes our review. In order to do actual computations involving trigonometric functions and logarithms, you will need their numerical values. You can obtain these from a scientific calculator or from published tables such as those in the *The Handbook of Chemistry and Physics* (Chemical Rubber Publishing Co.). Also, scientific computers invariably have programs for generating these functions.

Before going on, there are a few features of this book you ought to know about. The last chapter, Chapter 4, summarizes the first three chapters to help you review what you have learned. Take a look at that summary if you feel the need. In addition, starting on page 245, there is a collection of review problems with answers. In addition to the index at the back of the book, there is a separate index of symbols on page 261.

As soon as you are ready, go to Chapter 2.

CHAPTER TWO
Differential Calculus

In this chapter you will learn

- What is meant by the *limit* of a function
- How the *derivative* of a function is defined
- How to interpret derivatives graphically
- Some shortcuts for finding derivatives
- How to recognize the derivatives of some common functions
- How to find the maximum or minimum values of functions
- How to apply differential calculus to a variety of problems

Limits

97

Before tackling differential calculus, we must learn about limits. The idea of a limit may be new to you, but it is at the heart of calculus and it is important to understand the material in this section before going on. Once you understand limits, you should be able to grasp the ideas of differential calculus quite readily.

Limits are so important in calculus that we will discuss them from two different points of view. First, we will discuss limits from an intuitive point of view. Then, we will give a precise mathematical definition.

Go to **98**.

98

Here is a little bit of mathematical shorthand which will be useful in this section.

Suppose a variable x has values lying in an interval with the following properties:

1. The interval surrounds some number a.
2. The difference between x and a is less than another number B.
3. x does not take the particular values a. (We will see later why this point is excluded.)

The above three statements can be summarized by the following:

$\lvert x - a \rvert > 0$	(This statement means x cannot have the value a.)
$\lvert x - a \rvert < B$	(The magnitude of the difference between x and a is less than B.)

These relations can be combined in the single statement:

$$0 < \lvert x - a \rvert < B.$$

(If you need to review the symbols used here, see frame **20**.)

The values of x which satisfy $0 < \lvert x - a \rvert < B$ are indicated by the interval along the x-axis shown in the figure.

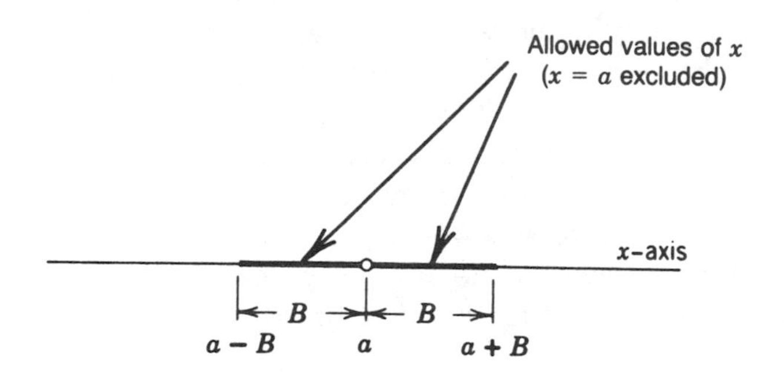

Go to **99**.

99

We begin our discussion of limits with an example. We are going to work with the equation $y = f(x) = x^2$, as shown in the graph. P is the point on corresponding to $x = 3$, $y = 9$.

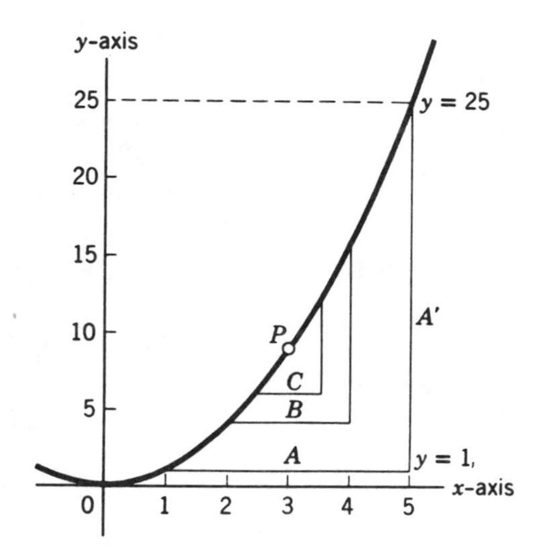

Let us concentrate on the behavior of y for values of x in an interval about $x = 3$. For reasons which we shall see shortly, it is important to exclude the particular point of interest P, and to remind us of this, the point is encircled on the curve.

We start by considering values of y corresponding to values of x in an interval about $x = 3$, lying between $x = 1$ and $x = 5$. With the notation of the last frame, this can be written as $0 < |x-3| < 2$. This interval for x is shown by line A in the figure. The corresponding interval for y is shown by line A' and includes points between $y = 1$ and $y = 25$, except $y = 9$.

A smaller interval for x is shown by line B. Here $0 < |x-3| < 1$, and the corresponding interval for y is $4 < y < 16$, with $y = 9$ excluded.

The interval for x shown by the line C is given by $0 < |x-3| < 0.5$. Write the corresponding interval for y in the blank below, assuming $y = 9$ is excluded.

____

In order to find the correct answer, go to **100**.

100

The interval for y which corresponds to $0 < |x - 3| < 0.5$ is

$$6.25 < y < 12.25$$

which you can check by substituting the values 2.5 and 3.5 for x in $y = x^2$ in order to find the values of y at either end point.

So far we have considered three successively smaller intervals of x about $x = 3$ and the corresponding intervals of y. Suppose we continue the process. The drawing shows the plot $y = x^2$ for values of x between 2.9 and 3.1. (This is an enlarged piece of the graph in frame **99**. Over the short distance shown the parabola looks practically straight.)

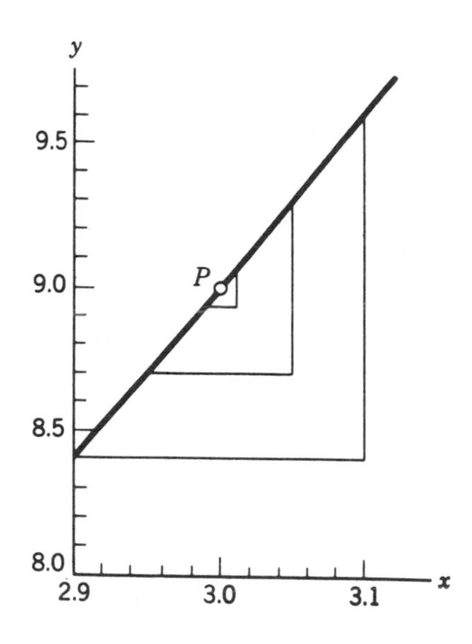

Three small intervals of x around $x = 3$ are shown along with the corresponding interval in y. The table below shows the values of y, corresponding to the boundaries of x at either end of the interval. (The last entry is for an interval too small to shown on the drawing.)

Interval of x	Corresponding interval of y
1–5	1–25
2–4	4–16
2.5–3.5	6.25–12.25
2.9–3.1	8.41–9.61
2.95–3.05	8.70–9.30
2.99–3.01	8.94–9.06
2.999–3.001	8.994–9.006

Go to **101**.

101

We hope it is apparent from the discussion in the last two frames that as we diminish the interval for x around $x = 3$, the values for $y = x^2$ cluster more and more closely about $y = 9$. In fact, it appears that we can make the values for y cluster as closely as we please about $y = 9$ by merely limiting x to a sufficiently small interval about $x = 3$. Because this is true, we say that the *limit* of x^2, as x approaches 3, is 9, and we write this

$$\lim_{x \to 3} x^2 = 9.$$

Let's put this in more general terms.

If a function $f(x)$ is defined for values of x about some fixed number a, and if, as x is confined to smaller and smaller intervals about a, the values of $f(x)$ cluster more and more closely about some specific number L, the number L is called the *limit* of $f(x)$ as x approaches a. The statement that "the limit of $f(x)$ as x approaches a is L" is customarily abbreviated by

$$\lim_{x \to a} f(x) = L.$$

In the example at the top of the page $f(x) = x^2$, $a = 3$, and $L = 9$.

The important idea in the definition is that the intervals we use lie around the point of interest a, but that the point itself is not included. In fact, $f(a)$, the value of the function at a, may be entirely different from $\lim_{x \to a} f(x)$, as we shall see.

Go to **102**.

102

You may be wondering why we have been giving such a complicated discussion of an apparently simple problem. Why bother with $\lim_{x \to 3} x^2 = 9$ when it is obvious that $x^2 = 9$ for $x = 3$?

The reason is that often the value of a function for a particular $x = a$ is not defined, whereas the limit as x approaches a is perfectly well defined. For instance at $\theta = 0$ the function $\frac{\sin \theta}{\theta}$ has the value $\frac{0}{0}$, which is meaningless. When we get to frame **110** we shall see that

$$\lim_{\theta \to 0} \frac{\sin \theta}{\theta} = 1.$$

As another illustration consider

$$f(x) = \frac{x^2 - 1}{x - 1}.$$

For $x = 1$, $f(1) = \frac{1 - 1}{1 - 1} = \frac{0}{0}$, which is not defined. However we can divide by $x - 1$ *provided* x is not equal to 1, and we obtain

$$f(x) = \frac{x^2 - 1}{x - 1} = \frac{(x + 1)(x - 1)}{x - 1} = x + 1.$$

Therefore, even though $f(1)$ is not defined,

$$\lim_{x \to 1} f(x) = \lim_{x \to 1} (x + 1) = 2.$$

Formal justification of the last two steps is given in Appendix A1, along with a number of rules for handling limits. There is no need to read the appendix now unless you are really interested.

We could also have obtained the above result graphically by studying the graph of the function in the neighborhood of $x = 1$ as we did in frame **99**.

Go to **103**.

103

To see whether you have caught on, find the limit of the following slightly more complicated functions by procedures similar to the above. (You will probably have to work these out on paper. Both of them involve a little algebraic manipulation.)

(a) $\lim_{x \to 0} \frac{(1 + x)^2 - 1}{x} =$ [1 | x | −1 | 2]

(b) $\lim_{x \to 0} \frac{1 - (1 + x)^3}{x} =$ [1 | x | 3 | −3]

If right, go to **105**.
Otherwise, go to **104**.

104

Here are the solutions to the problems in **103**:

(a) $$\lim_{x\to 0}\frac{(1+x)^2-1}{x} = \lim_{x\to 0}\frac{(1+2x+x^2)-1}{x}$$

$$= \lim_{x\to 0}\frac{2x+x^2}{x} = \lim_{x\to 0}(2+x) = \lim_{x\to 0} 2 + \lim_{x\to 0} x = 2.$$

(b) $$\lim_{x\to 0}\frac{1-(1+x)^3}{x} = \lim_{x\to 0}\frac{1-(1+x)(1+x)(1+x)}{x}$$

$$= \lim_{x\to 0}\frac{1-(1+3x+3x^2+x^3)}{x} \lim_{x\to 0}(-3-3x-x^2)$$

$$= \lim_{x\to 0}(-3) + \lim_{x\to 0}(-3x) + \lim_{x\to 0}(-x^2) = -3.$$

Again, if you would like justification of the steps used in these solutions, see Appendix A2.

Go to **105**.

105

So far we have discussed limits using expressions such as "confined to a smaller and smaller interval" and "clustering more and more closely." These expressions convey the intuitive meaning of a limit, but they are not precise mathematical statements. Now we are ready for a precise definition of a limit. [Since it is an almost universal custom, in the definition of a limit we will use the Greek letters δ (delta) and ϵ (epsilon).]

Definition of a Limit

Let $f(x)$ be defined for all x in an interval about $x = a$, but not necessarily at $x = a$. If there is a number L such that to each positive number ϵ there corresponds a positive number δ such that

$$|f(x) - L| < \epsilon \qquad \text{provided } 0 < |x - a| < \delta,$$

we say that L is the limit of $f(x)$ as x approaches a, and write

$$\lim_{x\to a} f(x) = L.$$

To see how to apply this definition,

Go to **106**.

Answers: **(103)** 2, −3

106

The formal definition of a limit in frame **105** provides a clear basis for settling a dispute as to whether the limit exists and is L. Suppose we assert that $\lim_{x \to a} f(x) = L$, and an opponent disagrees. As a first step, we tell her to pick a positive number ϵ, as small as she pleases, say 0.001, or if she wants to be difficult, 10^{-100}. Our task is to find some other number δ, such that for all x in the interval $0 < | x - a | < \delta$, the difference between $f(x)$ and L is smaller than ϵ. If we can always do this, we win the argument—the limit exists and is L. These steps are illustrated for a particular function in the drawings below.

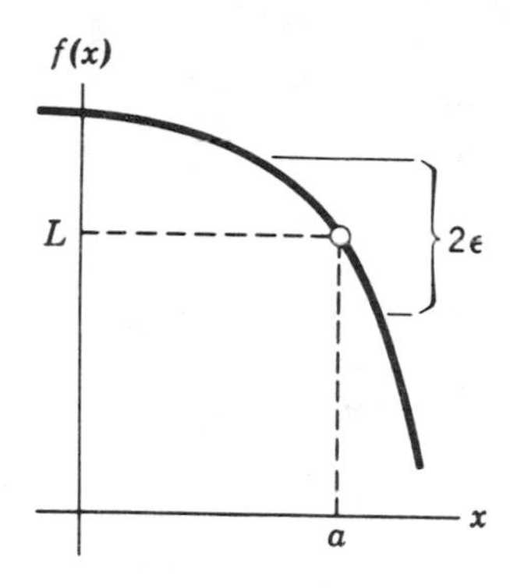

Our opponent has challenged us to find a δ to fit this ϵ.

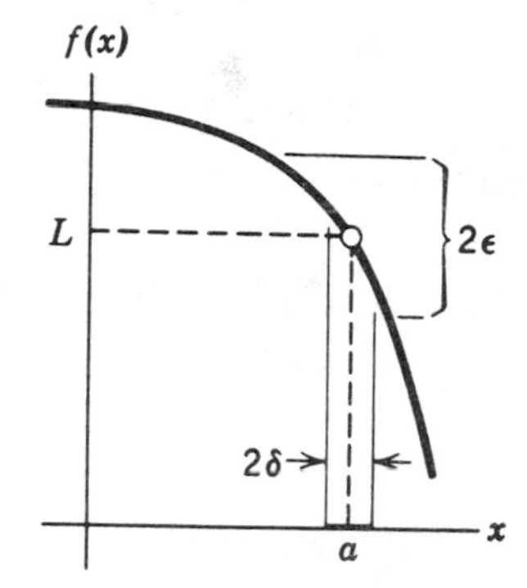

Here is one choice of δ. Obviously, for all values of x in the interval shown, $f(x)$ will satisfy $| f(x) - L | < \epsilon$.

It may be that our opponent can find an ϵ such that we can never find a δ, no matter how small, that satisfies our requirement. In this case, she wins and $f(x)$ does not have the limit L. (In frame **114** we will come to an example of a function which does not have a limit.)

Go to **107**.

107

In the examples we have studied so far, the function has been expressed by a single equation. However, this is not necessarily the case.

(continued)

Here is an example to show this.

$$f(x) = 1 \qquad \text{for } x \neq 2,$$
$$f(x) = 3 \qquad \text{for } x = 2.$$

(The symbol $\neq$ means "not equal.")
A suggestive sketch of this peculiar function is shown. You should be able to convince yourself that $\lim_{x\to 2} f(x) = 1$, whereas $f(2) = 3$.

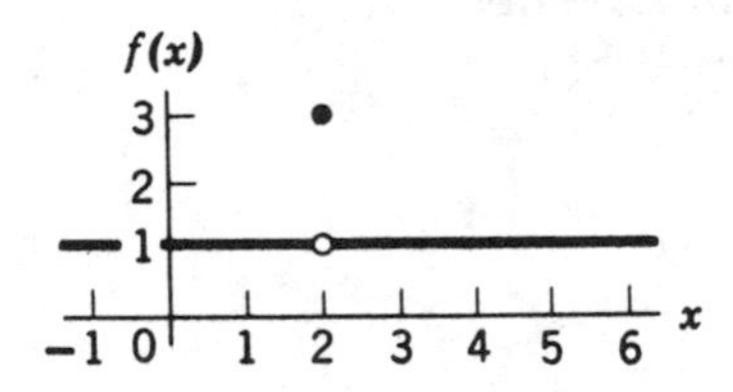

If you would like further explanation of this, go to **108**.

Otherwise, go to **109**.

108

For every value of x except $x = 2$, the value of $f(x) = 1$. Consequently, $f(x) - 1 = 0$ for all x except $x = 2$. Since 0 is less than the smallest positive number ϵ that your opponent could select, it follows from the definition of a limit that $\lim_{x\to 2} f(x) = 1$, even though $f(2) = 3$.

Go to **109**.

109

Here is another function which has a well-defined limit at a point but which can't be evaluated at that point: $f(x) = (1 + x)^{1/x}$. The value of $f(x)$ at $x = 0$ is quite puzzling. However, it is possible to find $\lim_{x\to 0} (1 + x)^{1/x}$.

Most scientific calculators have the function y^x. If you have such a calculator, determine the values in the following table

x	$(1 + x)^{1/x}$
1	
0.1	
0.01	
0.001	
0.0001	
0.00001	

The limit of $(1 + x)^{1/x}$ as $x \to 0$ will play an important role in our study of logarithms. It is given a special symbol, e. Like π, e is an unending and unrepeating decimal; it is irrational. The value of e is 2.7182828 . . .. If you tried evaluating e with a calculator, the last entry in the table should give correct values for the first four digits after the decimal point.

Go to **110**.

110

The actual procedure for finding a limit varies from problem to problem. There are a number of theorems for finding the limits of simple functions in Appendix A2, which you should read if you are interested. The result mentioned earlier,

$$\lim_{\theta \to 0} \frac{\sin \theta}{\theta} = 1$$

is proven in Appendix A3.

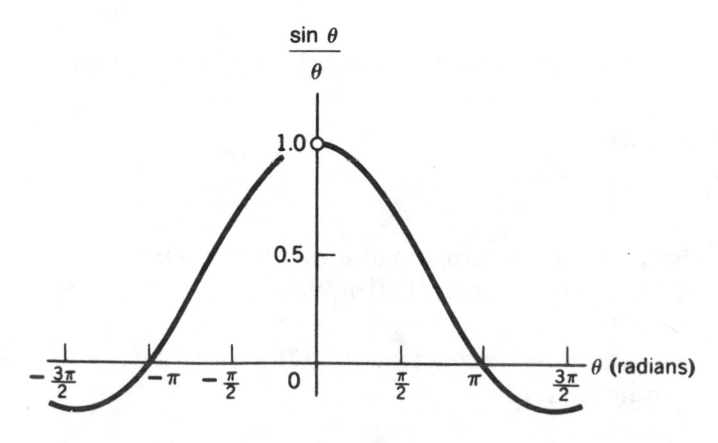

(continued)

You can see that this result is reasonable by graphing the function $\frac{\sin \theta}{\theta}$ as shown above. If you have a calculator, explore for yourself values of $\frac{\sin \theta}{\theta}$ as θ approaches zero. If you try to evaluate the function at $\theta = 0$, most calculators will indicate an error. This is as it should be, since the function is not defined at $\theta = 0$. Nevertheless, its limit is well defined and has the value 1.

Go to **111**.

111

So far in most of our discussion of limits we have been careful to exclude the actual value of $f(x)$ at the point of interest, a. In fact, $f(a)$ does not even need to be defined for the limit to exist (as in the last frame). However, frequently $f(a)$ is defined. If this is so, and if in addition

$$\lim_{x \to 0} f(x) = f(a),$$

then the function is said to be *continuous* at a. To summarize, fill in the blanks:

A function f is continuous at $x = a$ if

1. $f(a)$ is ____________________.
2. $\lim_{x \to a} f(x) =$ __________.

Check your answers in frame **112**.

112

Here are the correct answers: A function f is continuous at $x = a$ if

1. $f(a)$ is *defined*.
2. $\lim_{x \to a} f(x) = f(a)$.

A more picturesque description of a continuous function is that it is a function you can graph without lifting your pencil from the paper in the region of interest.

Try to determine whether each of the following functions is continuous at the point indicated.

1. $f(x) = \dfrac{x^2 + 3}{9 - x^2}$.

   At $x = 3$, $f(x)$ is [continuous | discontinuous]

2. $f(x) \begin{cases} 1, & x \geqslant 0, \\ 0, & x < 0. \end{cases}$

   At $x = 1$, $f(x)$ is [continuous | discontinuous]

3. $f(x) = |x|$.

   At $x = 0$, $f(x)$ is [continuous | discontinuous]

4. $f(x) = \dfrac{\sin x}{x}$.

   At $x = 0$, $f(x)$ is [continuous | discontinuous]

If you made any mistakes, or want more explanation, go to **113**.
Otherwise, skip on to **114**.

113

Here are the explanations of the problems in frame **112**.

1. At $x = 3$, $f(x) = \dfrac{x^2 + 3}{9 - x^2} = \dfrac{12}{0}$. This is an undefined expression and, therefore, the function is not continuous at $x = 3$.
2. Here is a plot of the function given.

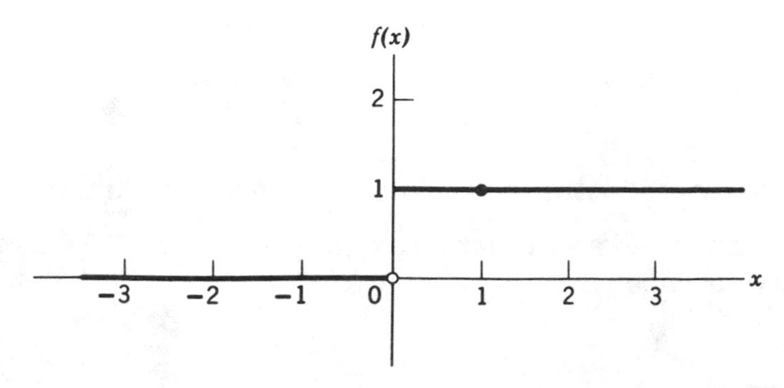

(continued)

This function satisfies both conditions for continuity at $x = 1$, and is thus continuous there. (It is, however, discontinuous at $x = 0$.)

3. Here is a plot of $f(x) = |x|$.

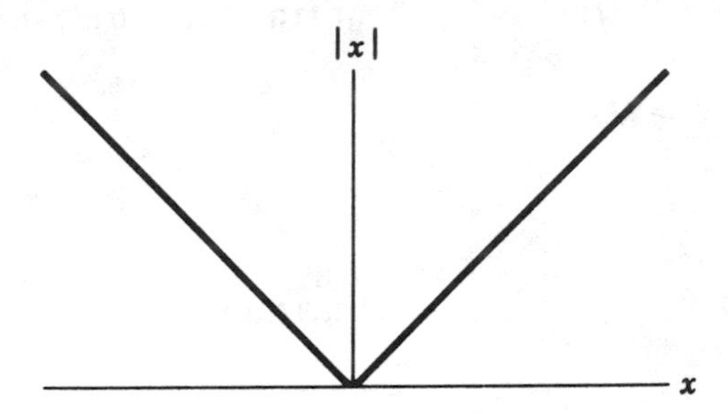

This function is continuous at $x = 0$ since it satisfies all the formal requirements.

4. As discussed in frame **110**, $\frac{\sin x}{x}$ is not defined at $x = 0$. (It is, however, continuous for all other values of x.)

Go to **114**.

114

Before leaving the subject of limits, it is worth looking at some examples of functions which somewhere have no limit. One such function is that described in problem 2 of the last frame. The graph of the function is shown in the figure. We can show that this function has no limit at $x = 0$ by following the procedure described in the definition of a limit.

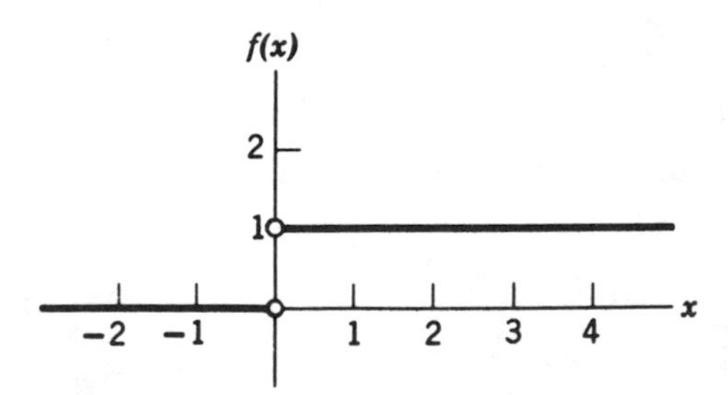

For purposes of illustration, suppose we guess that $\lim_{x \to 0} f(x) = 1$. Next, our opponent chooses a value for ϵ, say $\frac{1}{4}$. Now, for $|x - 0| < \delta$, where δ is *any* positive number,

---

Answers: (112) (1) discontinuous, (2) continuous, (3) continuous, (4) discontinuous

$$| f(x) - 1 | = \begin{cases} | 1 - 1 | = 0 & \text{if } x > 0, \\ | 0 - 1 | = 1 & \text{if } x < 0. \end{cases}$$

Therefore, for all negative values of x in the interval, $| f(x) - 1 | = 1$, which is greater than $\epsilon = 1/4$. Thus 1 is not the limit. You should be able to convince yourself that there is *no* number L which satisfies the criterion since $f(x)$ changes by 1 when x goes from negative to positive values.

Go to **115**.

115

Here is another example of a function which has no limit at a point. From the graph it is obvious that $\cot \theta$ has no limit as $\theta \to 0$. Instead of clustering more and more closely to any number, L, the value of the function gets increasingly larger as $\theta \to 0$ in the direction shown by A, and increasingly more negative as $\theta \to 0$ in the direction shown by B.

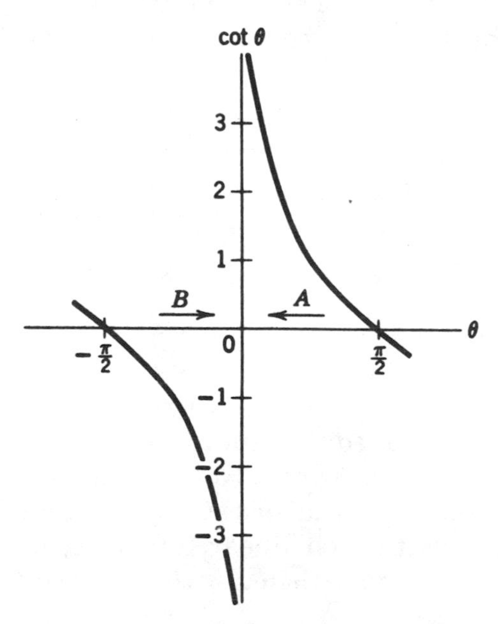

This concludes our study of the limit of a function for the present. If you would like some more practice with limits, see review problems 21 through 28 on page 246.

Now we are ready to go on to the next section.

Go to **116**.

Velocity

116

We have been getting a little abstract, so before we go on to differential calculus, let's talk about something down to earth: motion. As a matter of fact, Leibniz and Newton invented calculus because they were concerned with problems of motion, so it is a good place to start. Besides, you already know quite a bit about motion.

Go to **117**.

117

Here is a warm-up problem. In this chapter the motion is always along a straight line.

A train travels away at a velocity v mph (miles per hour). At $t = 0$, it is distance S_0 from us. (The subscript on S_0 is to avoid confusion. S_0 is a particular distance and is a constant; S is a variable.) Write the equation for the distance the train is from us S in terms of time t. (Take the unit of t to be hours.)

$$S = \_\_\_\_\_\_\_\_.$$

Go to **118** for the answer to this.

118

If you wrote $S = S_0 + vt$, you are correct. Go on to frame **119**.

If your answer was not equivalent to the above, you should convince yourself that this answer is correct. Note that it yields $S = S_0$ when $t = 0$, as required. The equation is that of a straight line, and it might be a good idea to review the section on linear functions, frames **23–39**, before continuing. When you are satisfied with this result,

Go to **119**.

119

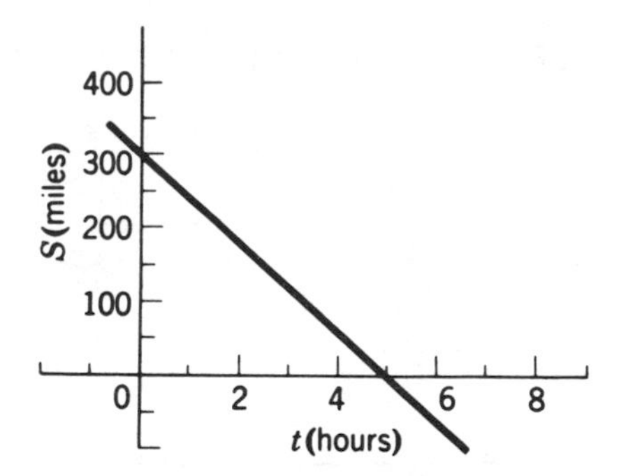

Here is a plot of the positions at different times of a train going in a straight line. Obviously, this represents a linear equation. Write the equation for the position of this train (in miles) in terms of time (in hours).

$$S = \_\_\_\_\_\_\_\_\_\_\_\_\_\_\_\_.$$

Find the velocity of the train from your equation.

$$v = \_\_\_\_\_\_\_\_\_\_\_\_\_\_\_\_.$$

Go to **120** for the correct answers.

120

Here are the answers to the questions in frame **119**.

$$S = -60t + 300 \text{ miles},$$

$$v = -60 \text{ mph}.$$

The velocity is negative because S decreases with increasing time. (Note that the velocity along a straight line is positive or negative depending on the direction of motion. The *speed*, which is the magnitude of the velocity, is always positive.) If you would like further discussion, review frames **33** and **34**.

Go to **121**.

121

Here is another plot of position of a train traveling in a straight line.

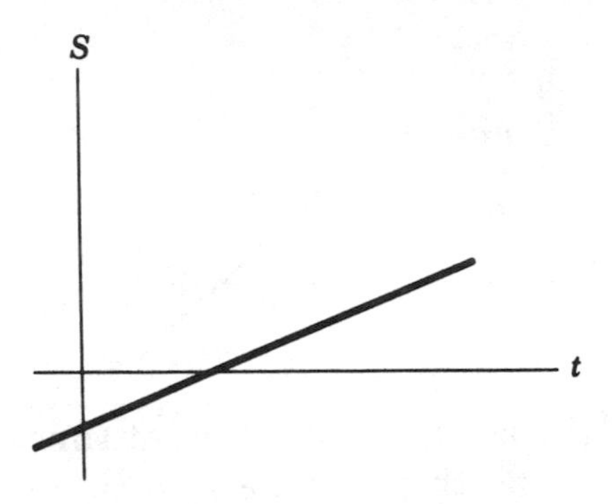

The property of the line which represents the *velocity* of the train is the ______________ of the line.

Go to **122** for the answer.

122

The property of the line which represents the velocity of the train is the *slope* of the line.

If you wrote this, go right on to **123**.

If you wrote anything else, or nothing at all, then you may have forgotten what we reviewed back in frames **23–39**. You should go over that section once again (particularly frames **33** and **34**) and think about this problem before going on. At least convince yourself that the slope really represents the velocity.

Go to **123**.

123

On the next page are plots of the positions vs. time of six objects moving along straight lines. Which plot corresponds to the object that

Has the greatest velocity forward? [*a* | *b* | *c* | *d* | *e* | *f*]

Is moving backward most rapidly? [*a* | *b* | *c* | *d* | *e* | *f*]

Is at rest? [*a* | *b* | *c* | *d* | *e* | *f*]

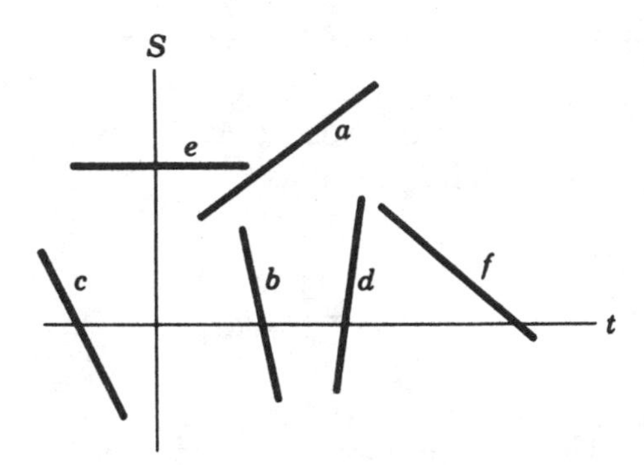

If all right, go to **125**.
If any wrong, go to **124**.

124

The velocity of the object is given by the slope of the plot of its distance against time. Don't confuse the slope of a line with its location.

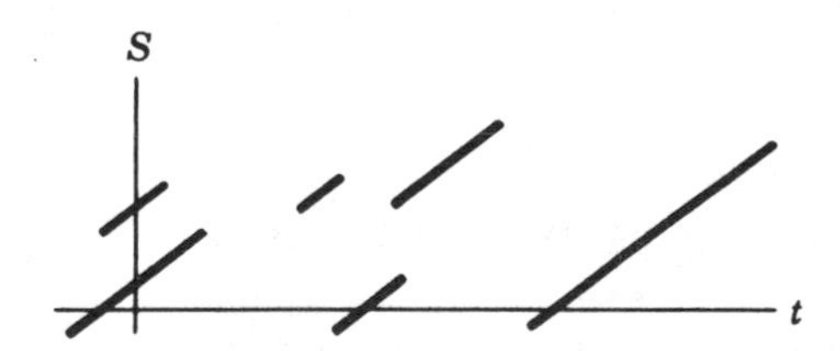

All the above lines have the same slope.

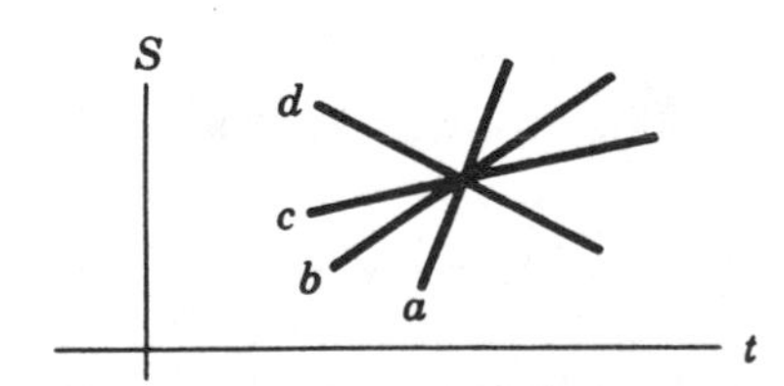

All these lines have different slopes.

A positive slope means that distance is increasing with time, which corresponds to a positive velocity. Likewise, a negative slope means that distance is decreasing in time, which means the velocity is negative. If you need to review the idea of slope, look at frames **25–27** before continuing.

Which line in the figure above on the right has

Negative slope? [*a* | *b* | *c* | *d*]

Greatest positive slope? [*a* | *b* | *c* | *d*]

Go to **125**.

125

So far, the velocities we have considered have all been constant in time. But what if the velocity changes?

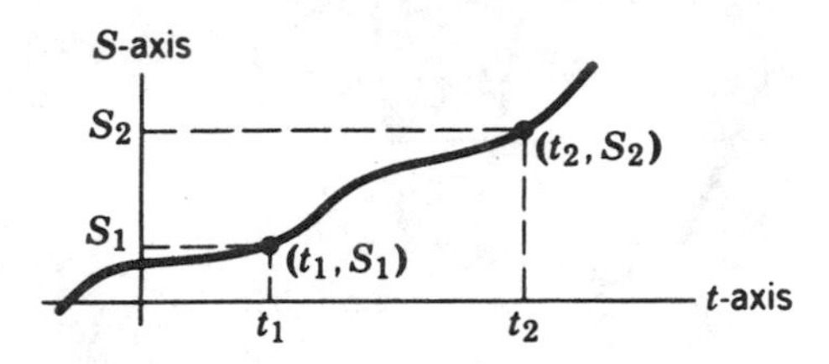

Here is a plot of the position of a car which is traveling with varying velocity along a straight line. In order to describe this motion, we introduce the *average velocity* $\bar{v}$ (read as "v bar"), which is the ratio of the net distance traveled to the time taken. For example, between the times t_1 and t_2 the car went a net distance $S_2 - S_1$, so $(S_2 - S_1)/(t_2 - t_1)$ was its ________________________ during the time.

Go to **126**.

126

The answer to frame **125** is

$(S_2 - S_1)/(t_2 - t_1)$ was its *average velocity* during the time.

(The single word "velocity" is not a correct answer.)

Go to **127**.

127

In addition to defining the average velocity $\bar{v}$ algebraically,

$$\bar{v} = \frac{S_2 - S_1}{t_2 - t_1},$$

we can interpret $\bar{v}$ graphically. If we draw a straight line between the points (t_1, S_1) and (t_2, S_2), then the average velocity is simply the *slope* of that line.

Answers: (123) *d, b, e*
(124) *d, a*

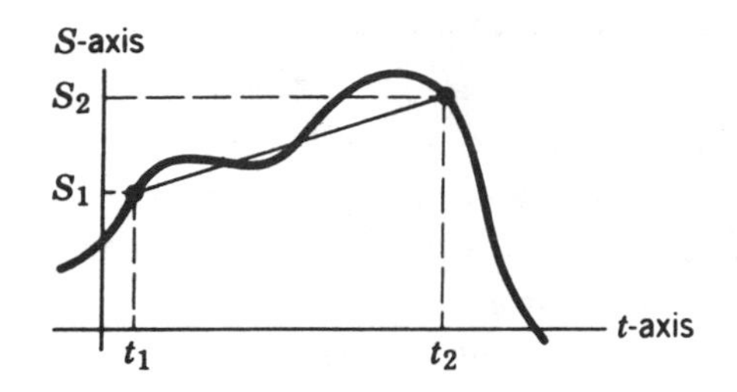

Go to **128**.

128

During which interval was the average velocity

Closest to 0? [1 | 2 | 3]

Largest forward? [1 | 2 | 3]

Largest backward? [1 | 2 | 3]

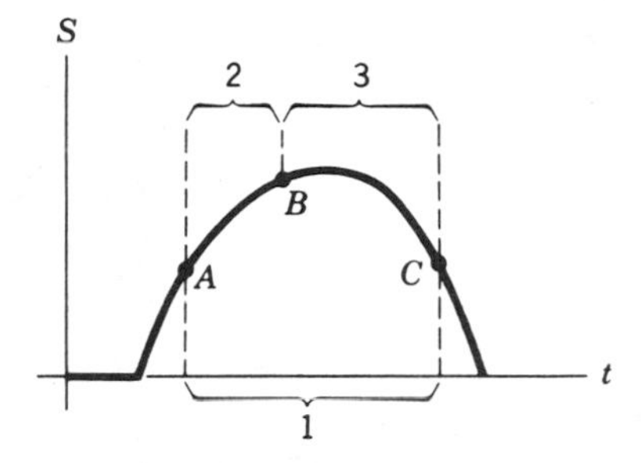

If right, go to **130**.
If wrong, go to **129**.

129

Since you missed the last problem, we'll analyze it in detail.

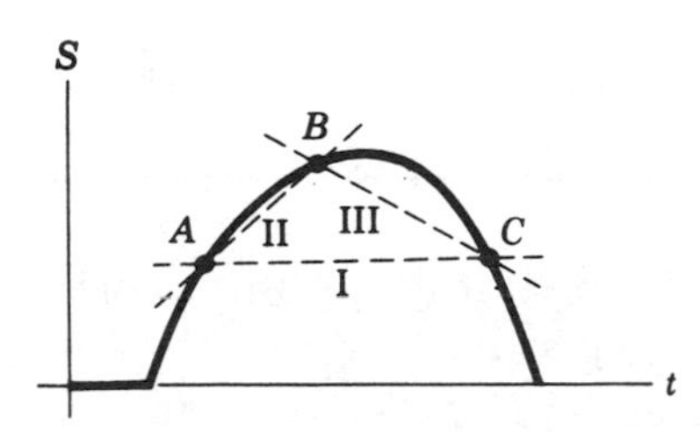

Here are straight lines drawn through the points A, B, C. Line I has a very small slope and corresponds to almost 0 velocity. Line II has positive slope, and line III has negative slope, corresponding to positive and negative average velocities, respectively.

Go to **130**.

130

We now extend our idea of velocity in a very important manner: instead of asking "what is the average velocity **between** time t_1 and t_2?" let us ask "what is the velocity **at** time t_1?" The velocity at a particular time is called the **instantaneous** velocity. This is a new term, and we will give it a precise definition shortly even though it may already be somewhat familiar to you.

Go to **131**.

131

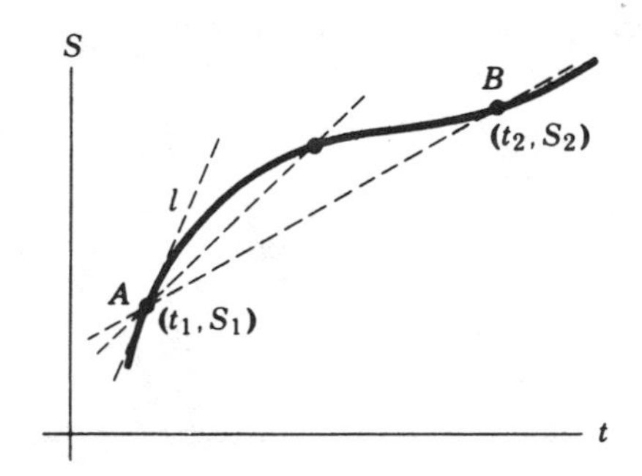

We can give a graphical meaning to the idea of instantaneous velocity. The average velocity is the slope of a straight line joining two points on the curve, (t_1, S_1) and (t_2, S_2). To find the instantaneous velocity, we want t_2 to be very close to t_1. As we let point B on the curve approach point A (i.e., as we consider intervals of time, starting at t_1 which become shorter and shorter), the slope of the line joining A and B approaches the slope of the line which is labeled l. The instantaneous velocity is then the *slope* of line l. In a sense, then, the straight line l has the same slope as the curve at the point A. Line l is called a *tangent* to the curve.

Go to **132**.

132

Here is where the idea of a limit becomes very important. If we draw a straight line through the given point A on the curve and some other point on the curve B, and then let B get closer and closer to A, the slope of the straight line approaches a unique value and can be identified with the *slope* of the curve at A. What we must do is consider the *limit* of the slope of the line through A and B as $B \to A$.

Now, go to **133**.

Answers: (128) 1, 2, 3

133

We will now give a precise meaning to the intuitive idea of instantaneous velocity as the slope of a curve at a point. We start by considering the average velocity:

$\bar{v} = (S_2 - S_1)/(t_2 - t_1)$ = the slope of the line connecting points 1 and 2.

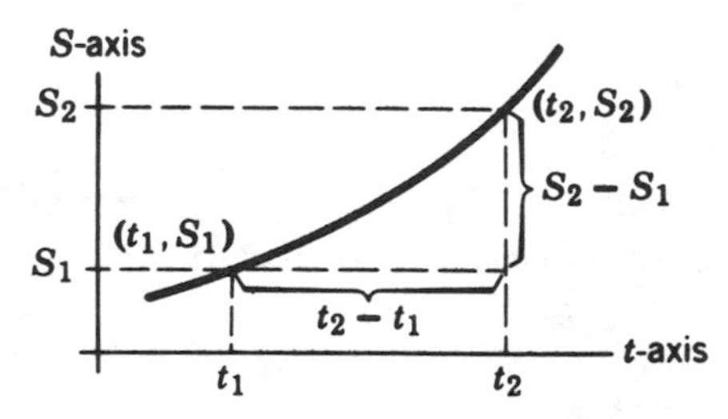

As $t_2 \rightarrow t_1$, the average velocity approaches the instantaneous velocity, that is, $\bar{v} \rightarrow v$ as $t_2 \rightarrow t_1$, or

$$v = \lim_{t_2 \rightarrow t_1} \frac{S_2 - S_1}{t_2 - t_1}.$$

Go to **134**.

134

Since the ideas presented in the last few frames are very important, let's summarize them.

If a point moves from S_1 to S_2 during the time t_1 to t_2, then

$$(S_2 - S_1)/(t_2 - t_1)$$

is the __________ __________, $\bar{v}$.

If we consider the limit of the average velocity as the averaging time goes to zero, the result is called the __________ __________, v.

Now let's try to present these ideas in a neater form. If you can, write a formal definition of v in the blank space.

v = __________

Go to frame **135** for the answers.

135

The correct answers to frame **134** are the following:
If a point moves from S_1 to S_2 during the time t_1 to t_2, then

$$(S_2 - S_1)/(t_2 - t_1)$$

is the *average velocity* $\bar{v}$.

If we consider the limit of the average velocity as the averaging time goes to zero, the result is called the *instantaneous velocity* v.

$$v = \lim_{t \to t_1} \frac{S_2 - S_1}{t_2 - t_1}.$$

If you wrote this, congratulations! Go on to **136**.

If you wrote something different, go back to frame **133** and work your way to this frame once more.

Then go on to **136**.

136

The Greek capital Δ ("delta") is often used to indicate the change in a variable. Thus, to make the notation more succinct, we can write $\Delta S = S_2 - S_1$, and $\Delta t = t_2 - t_1$. (ΔS is a single symbol read as "delta S"; it does not mean $\Delta \times S$.) Although this notation may be new, it is worth the effort to get used to it since it saves lots of writing.

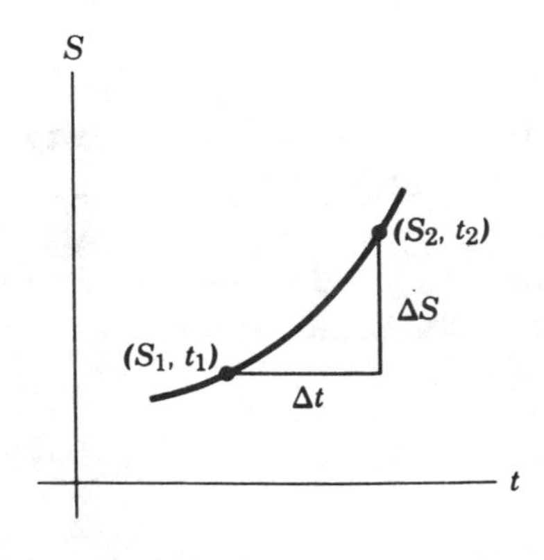

With this notation, our definition of instantaneous velocity is

$$v = \underline{\hspace{4cm}}$$

Go to **137** to find the correct answer.

137

If you wrote

$$v = \lim_{\Delta t \to 0} \frac{\Delta S}{\Delta t},$$

you are really catching on. Go ahead to frame **138**.

If you missed this, study frames **134–136** before going to **138**.

138

Now we are going to calculate an instantaneous velocity by analyzing an example step by step. Later on we will find shortcuts for doing this.

Suppose that we are given the following expression relating position and time:

$$S = f(t) = kt^2 \qquad (k \text{ is a constant}).$$

The goal is to find $\Delta S = f(t + \Delta t) - f(t)$, for any Δt, and then to evaluate the limit $\Delta S/\Delta t$ as $\Delta t \to 0$.

Here are the steps

$$\begin{aligned} \Delta S = f(t + \Delta t) - f(t) &= k(t + \Delta t)^2 - kt^2 \\ &= k[t^2 + 2t\,\Delta t + (\Delta t)^2] - kt^2 \\ &= k[2t\,\Delta t + (\Delta t)^2], \end{aligned}$$

$$\frac{\Delta S}{\Delta t} = \frac{k[2t\,\Delta t + (\Delta t)^2]}{\Delta t} = 2kt + k\,\Delta t,$$

$$v = \lim_{\Delta t \to 0} \frac{\Delta S}{\Delta t} = \lim_{\Delta t \to 0} (2kt + k\,\Delta t) = 2kt.$$

A simpler problem for you to try is in the next frame.

Go to **139**.

139

Suppose we are given that $S = f(t) = v_0 t + S_0$. The problem is to find the instantaneous velocity from our definition.

In time Δt the point moves distance ΔS.

$$\Delta S = \underline{\hspace{6cm}}.$$

$$v = \lim_{\Delta t \to 0} \frac{\Delta S}{\Delta t} = \underline{\hspace{4cm}}.$$

Write in the answers and go to **140**.

140

If you wrote

$$\Delta S = v_0 \, \Delta t$$

and

$$v = \lim_{\Delta t \to 0} \frac{\Delta S}{\Delta t} = v_0,$$

you are correct and can skip on to frame **142**.

If you wrote something different, study the detailed explanation in frame **141**.

141

Here is the correct procedure. Since $S = f(t) = v_0 t + S_0$,

$$\begin{aligned} \Delta S &= f(t + \Delta t) - f(t) \\ &= v_0 \, (t + \Delta t) + S_0 - (v_0 t + S_0) \\ &= v_0 \, \Delta t, \end{aligned}$$

$$\lim_{\Delta t \to 0} \frac{\Delta S}{\Delta t} = \lim_{\Delta tx \to 0} \frac{v_0 \, \Delta t}{\Delta t} = \lim_{\Delta t \to 0} v_0 = v_0.$$

The instantaneous velocity and the average velocity are the same in this case, since the velocity is a constant, v_0.

Go to frame **142**.

142

Here is a problem for you to work out. Suppose the position of an object is given by

$$S = f(t) = kt^2 + lt + S_0,$$

where k, l, and S_0 are constants. Find v.

$$v = \lim_{\Delta t \to 0} \frac{\Delta S}{\Delta t} = \_\_\_\_\_\_\_\_\_\_.$$

To check your answer, go to **143**.

143

The answer is

$$v = 2kt + l.$$

If you obtained this result, go on to frame **146**. Otherwise,

Go to **144**.

144

Here is the solution to the problem in frame **142**.

$$\begin{aligned} f(t) &= kt^2 + lt + S_0, \\ f(t + \Delta t) &= k(t + \Delta t)^2 + l(t + \Delta t) + S_0 \\ &= k[t^2 + 2t\,\Delta t + (\Delta t)^2] + l(t + \Delta t) + S_0, \\ \Delta S &= f(t + \Delta t) - f(t) = k[2t\,\Delta t + (\Delta t)^2] + l\,\Delta t, \\ v &= \lim_{\Delta t \to 0} \frac{\Delta S}{\Delta t} = \lim_{\Delta t \to 0} \left\{ \frac{k[2t\,\Delta t + (\Delta t)^2] + l\,\Delta t}{\Delta t} \right\} \\ &= \lim_{\Delta t \to 0} [k(2t + \Delta t) + l] = 2kt + l. \end{aligned}$$

(continued)

Now try this problem:

If $S = At^3$, where A is a constant, find v.

Answer: ______________

To check your solution, go to **145**.

145

Here is the answer: $v = 3At^2$. Go right on to frame **146** unless you would like to see the solution, in which case continue here.

$$S = At^3,$$
$$\Delta S = A(t + \Delta t)^3 - At^3$$
$$= A[t^3 + 3t^2\,\Delta t + 3t(\Delta t)^2 + (\Delta t)^3] - At^3$$
$$= 3At^2\,\Delta t + 3At(\Delta t)^2 + A(\Delta t)^3,$$
$$v = \lim_{\Delta t \to 0} \frac{\Delta S}{\Delta t} = \lim_{\Delta t \to 0} [3At^2 + 3At\,\Delta t + A(\Delta t)^2] = 3At^2.$$

Go to frame **146**.

Derivatives

146

In this section we will generalize our results on velocity. This will lead us to the idea of the *derivative* of a function, which is at the very heart of differential calculus.

Go to **147**.

147

Fill in the blanks below.
When we write $S = f(t)$, we are stating that position depends on time.

Here position is the dependent variable and time is the __________ variable.

The velocity is the rate of change of position with respect to time. By this we mean that velocity is (give the formal definition again):

$$v = \underline{\hspace{4cm}}$$

Go to frame **148** for the correct answers.

148

In the last frame you should have written

. . . time is the *independent* variable,

and

$$v = \lim_{\Delta t \to 0} \frac{\Delta S}{\Delta t}$$

In any case, go on to **149**.

149

Let us consider any continuous function defined by, say, $y = f(x)$. Now y is our dependent variable, and x is our independent variable. If we ask "At what rate does y change as x changes?", we can find the answer by taking the following limit:

$$\text{Rate of change of } y \text{ with respect to } x = \lim_{\Delta x \to 0} \frac{\Delta y}{\Delta x}.$$

Go to **150**.

150

You can give a geometrical meaning to $\lim_{\Delta x \to 0} \frac{\Delta y}{\Delta x}$, where $y = f(x)$. To do so, fill in the blanks. Geometrically, $\lim_{\Delta x \to 0} \frac{\Delta y}{\Delta x}$ can be found by drawing a straight line through the point (x, y) and the point (_______, _______)

(continued)

as shown. The slope of that line is given by $\frac{\Delta y}{\Delta x}$, and $\lim_{\Delta x \to 0} \frac{\Delta y}{\Delta x}$ is the ________________________ of the tangent to the curve at (x, y).

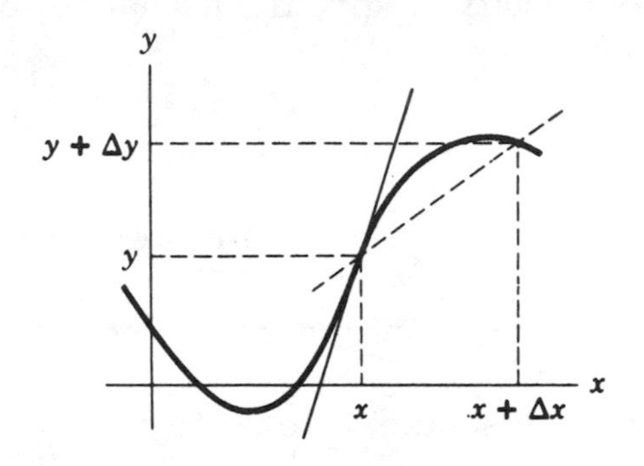

Go to **151**.

151

The correct insertions for frame **150** are

$$(x + \Delta x, y + \Delta y),$$

$\lim_{\Delta x \to} \frac{\Delta y}{\Delta x}$ is the slope of the tangent curve at (x, y). For brevity, the slope of the tangent to a curve is usually called *the slope of the curve.*

(If you would like to see a discussion of this, review frame **131** before continuing.)

Go to **152**.

152

Another way of writing $\frac{\Delta y}{\Delta x}$ is

$$\frac{y_2 - y_1}{x_2 - x_1} \quad \text{or} \quad \frac{f(x_2) - f(x_1)}{x_2 - x_1}.$$

If the notation used here still seems unfamiliar, review frame **136** before proceeding.

Go to **153**.

153

Let's review just once more.

If we want to know how y changes as x changes, we find out by calculating the following limit:

____

Fill in the blank and go on to **154**.

154

The correct answer to frame **153** is

$$\lim_{\Delta x \to 0} \frac{\Delta y}{\Delta x} \qquad \text{or} \qquad \lim_{x_2 \to x_1} \frac{y_2 - y_1}{x_2 - x_1}.$$

If you were correct go on to **155**.
If you missed this, go back to **149**.

155

Because the quantity $\lim_{\Delta x \to 0} \frac{\Delta y}{\Delta x}$ is so useful, we give it a special name and a special symbol.

$\lim_{\Delta x \to 0} \frac{\Delta y}{\Delta x}$ is called the *derivative* of y with respect to x, and it is often written with the symbol $\frac{dy}{dx}$.

$$\boxed{\frac{dy}{dx} = \lim_{\Delta x \to 0} \frac{\Delta y}{\Delta x},}$$

where $\Delta y = y(x + \Delta x) - y(x)$.

Once again: $\frac{dy}{dx}$ is the _______________ of _____ with respect to _____.

Go to **156** for the correct answer.

156

The correct answer is

$$\frac{dy}{dx} \text{ is the } \textit{derivative} \text{ of } y \text{ with respect to } x.$$

This symbol is read as "dee y by dee x." The derivative is frequently written in another form:

$$\frac{dy}{dx} = y'.$$

(The symbol y' is read as "y prime.") y' and $\frac{dy}{dx}$ mean the same thing:

$$y' = \frac{dy}{dx} = \lim_{\Delta x \to 0} \frac{\Delta y}{\Delta x}.$$

(Another symbol sometimes used for the derivative operator is D. Thus $Dy = y'$. However, we will not use the "D" symbol.)

Having two separate symbols for the derivative may look confusing at first, but they should both quickly become familiar. Each has its advantages. The symbol $\frac{dy}{dx}$ leaves no doubt that the independent variable is x, whereas y' might be ambiguous—y could be a function of some other variable, z. [To avoid such a confusion, the "prime" form is sometimes written as $y'(x)$.] On the other hand, the symbol $\frac{dy}{dx}$ can be cumbersome to write. More seriously, in the form $\frac{dy}{dx}$ the derivative looks like the simple ratio of two quantities, dy and dx, which it is not.

We can apply the idea of a derivative to the motion of velocity which we discussed earlier. Velocity is the rate of change of position with respect to time, so velocity is the *derivative* of position with respect to time.

Go to **157**.

157

Let's state the definition of a derivative using different variables. Suppose z is some independent variable, and q depends on z. Then the derivative of q with respect to z is

$$\frac{dq}{dz} = \underline{\hspace{4cm}}.$$

(Give formal definition.)

For the right answer, go to **158**.

158

Your answer should have been

$$\frac{dq}{dz} = \lim_{\Delta z \to 0} \frac{\Delta q}{\Delta z}.$$

If so, go to **159**.
If not, go back to frame **155** and try again.

159

The symbol $\frac{df}{dx}$ can be thought of as a derivative *operator* $\frac{d}{dx}$, operating on the function f.

If $f(x) = x^3 + 3$, then the derivative can be written in any of the following forms:

$$\frac{df}{dx} = \frac{d(x^3 + 3)}{dx} = \frac{d}{dx}(x^3 + 3).$$

Similarly,

$$\frac{d(\theta^2 \sin \theta)}{d\theta} = \frac{d}{d\theta}(\theta^2 \sin \theta).$$

(Here, θ is merely another variable.)

Thus $\frac{d}{dx}(\quad)$ means "differentiate with respect to x" whatever function $f(x)$ happens to be in the parentheses. In complete detail the symbol means that one should obtain an expression for

$$\Delta f = f(x + \Delta x) - f(x),$$

and then use it to evaluate

$$\frac{df}{dx} = \lim_{\Delta x \to 0} \frac{\Delta f}{\Delta x}.$$

However, as we shall see, one hardly ever goes through this formal limiting procedure to find a derivative. There are lots of shortcuts.

Go to **160**.

Graphs of Functions and Their Derivatives

160

We have just learned the formal definition of a derivative. Graphically, the derivative of a function $f(x)$ at some value of x is equivalent to the slope of a straight line which is tangent to the graph of the function at that point. Our chief concern in the rest of this chapter will be to find methods for evaluating derivatives of different functions. However, in doing this it is very helpful to have some sort of intuitive idea of how the derivative behaves, and we can obtain this by looking at the graph of the function. If the graph has a steep positive slope, the derivative is large and positive. If the graph has a slight slope downward, the derivative is small and negative. In this section we will get some practice putting to use such qualitative ideas as these, and in the following sections we will learn how to obtain derivatives precisely.

Go to **161**.

161

Here is a plot of the simple function $y = x$. At the top of the next page we have plotted $y' = \frac{dy}{dx}$. Since the slope of y is positive and constant, y' is a positive constant.

The graph indicates that $\frac{d}{dx}(x) = 1$. Can you prove this?

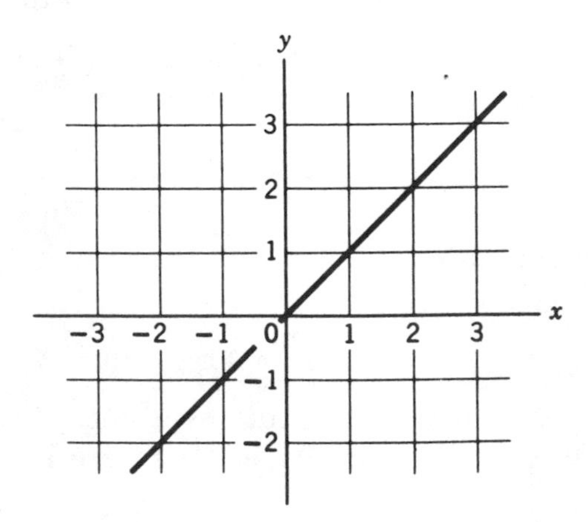

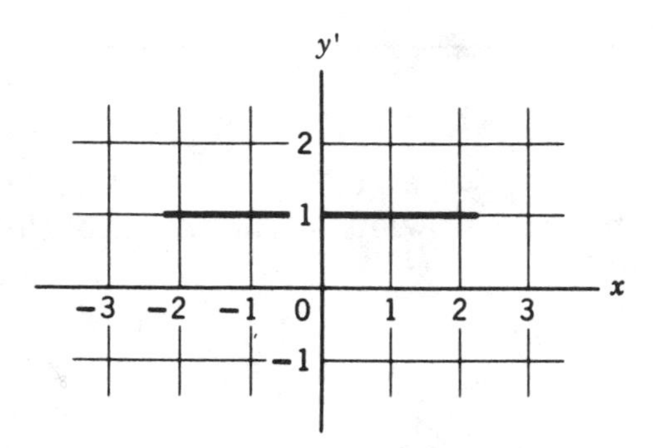

Go to **162**.

162

To prove that $\frac{d}{dx}(x) = 1$, let $y(x) = x$. Then

$$\Delta y = y(x + \Delta x) - y(x) = x + \Delta x - x = \Delta x.$$

Hence,

$$\frac{dy}{dx} = \lim_{\Delta x \to 0} \frac{\Delta y}{\Delta x} = \lim_{\Delta x \to 0} \frac{\Delta y}{\Delta x} = 1.$$

Here is a plot of $y = |x|$. (If you have forgotten the definition of $|x|$, see frame **20**.) On the coordinates below, sketch y'.

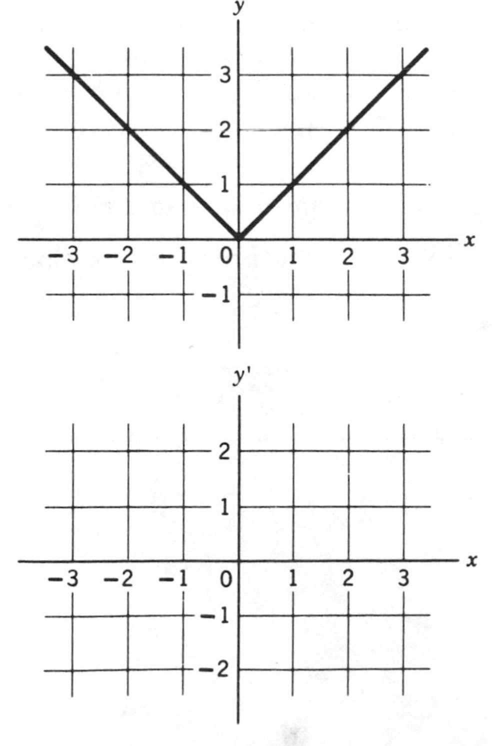

For the correct answer, go to **163**.

163

Here are sketches of $y = |\ x\ |$ and y'. If you drew this correctly, go on to **164**. If you made a mistake or want further explanation, continue here.

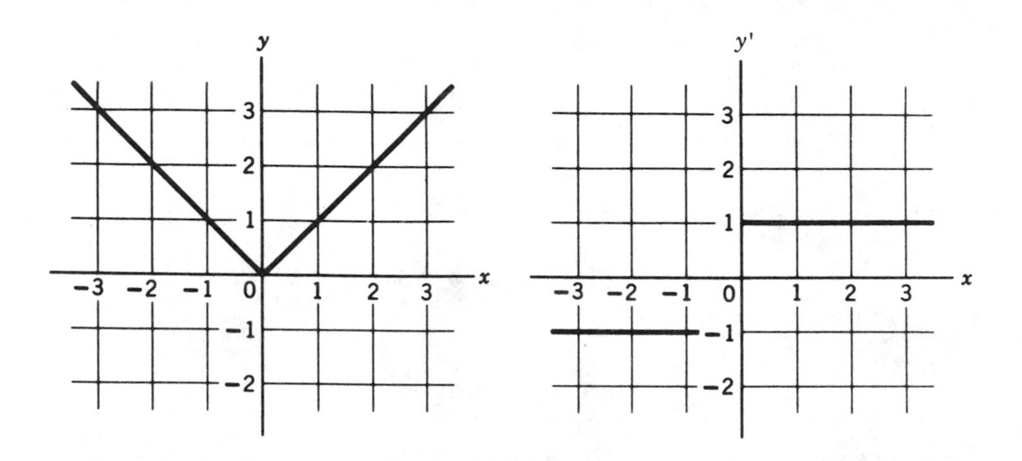

As you can see from the graph, $y = |\ x\ | = x$ for $x > 0$. So for $x > 0$ the problem is identical to that in frame **161**, and $y' = 1$. However, for $x < 0$, the slope of $|\ x\ |$ is negative and is easily seen to be -1. At $x = 0$, the slope is undefined, for it has the value $+1$ if we approach 0 along the positive x-axis and has the value -1 if we approach 0 along the negative x-axis.

Therefore, $\frac{d}{dx}(|\ x\ |)$ is discontinuous at $x = 0$. (The function x is continuous at this point, but the break in its slope at $x = 0$ causes a discontinuity in the derivative.)

Go to **164**.

164

Here is the graph of a function $y = f(x)$. Sketch its derivative in the space provided below. (The sketch does not need to be exact—just show the general features of y'.)

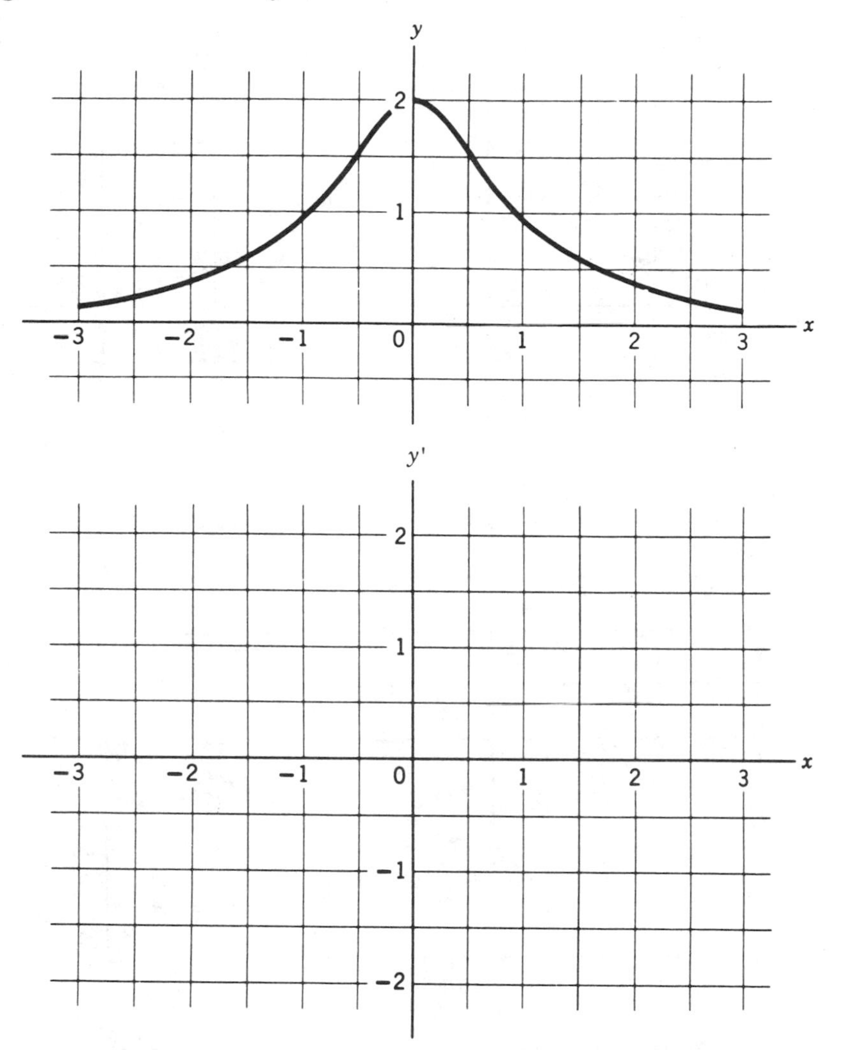

See **165** for the correct answer.

165

Here is the function and its derivative. If your sketch of y' is similar to that shown, go to **166**. Otherwise, read on.

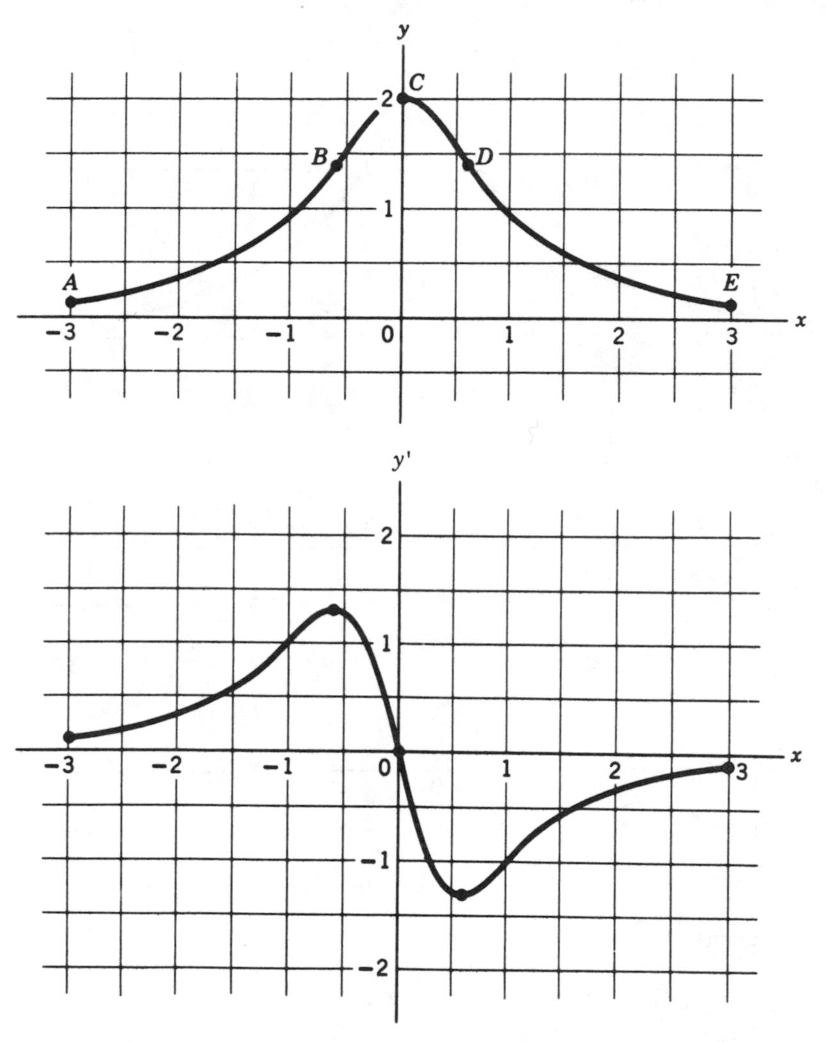

To see that the plot of y' is reasonable note that for $x < 0$, y increases with x so that y' is positive. The slope is greatest near point B, but it must abruptly decrease beyond B since it vanishes at C ($x = 0$). At D, y is decreasing rapidly, so y' is negative. At the extremes, A and E, the slope is small and y' is close to zero.

Go to **166**.

166

Let's look at the behavior of y' graphically for one more function. Here the plot of y and x is a semicircle. In the space below, make a rough sketch of y' for the interval illustrated.

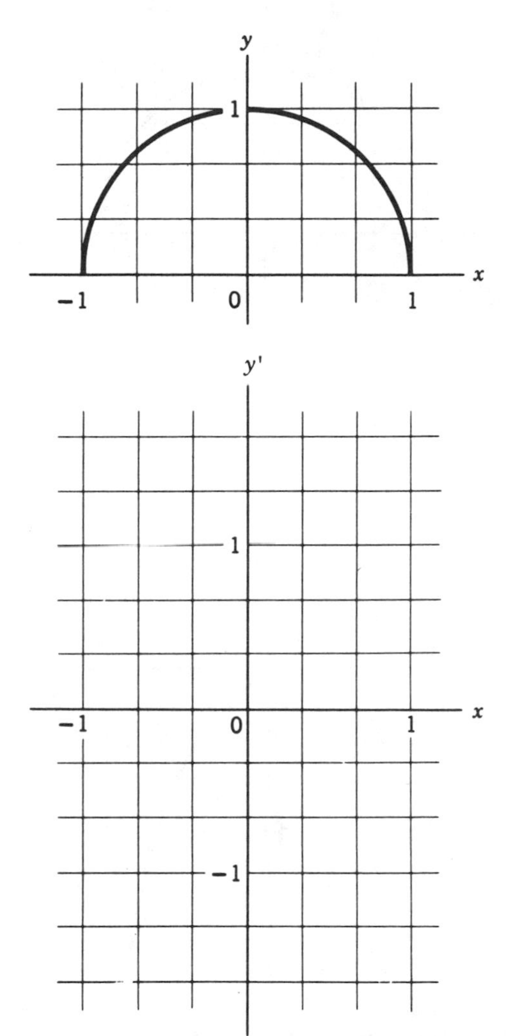

Go to **167** for the correct answer.

167

Here are the plots of y and y'. Read on if you would like further discussion of this. Otherwise, go to **168**.

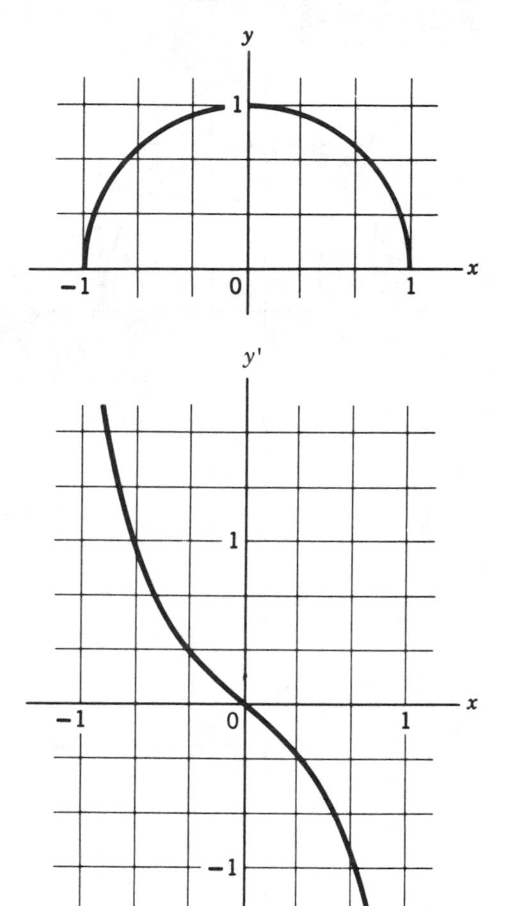

The slope of the semicircle does not behave nicely at the extreme values of x, so let's start by looking at $x = 0$. If we draw a line tangent to the curve at $x = 0$, it will be parallel to the x-axis, so the curve has 0 slope. Thus, $y' = 0$ at $x = 0$. For $x > 0$, a line tangent to the curve has negative slope, so $y' < 0$. As x approaches 1 the tangent becomes increasingly steep, and y' becomes increasingly negative. In fact, as $x \rightarrow 1, y' \rightarrow -\infty$.

From this discussion it should be easy to find y' for $x < 0$.

Go to **168**.

168

If you understand all the examples in this section, skip on to the next section. However, if you would like a little more practice, try sketching the derivatives for each function shown. The correct sketches are given in frame **169** without any discussion.

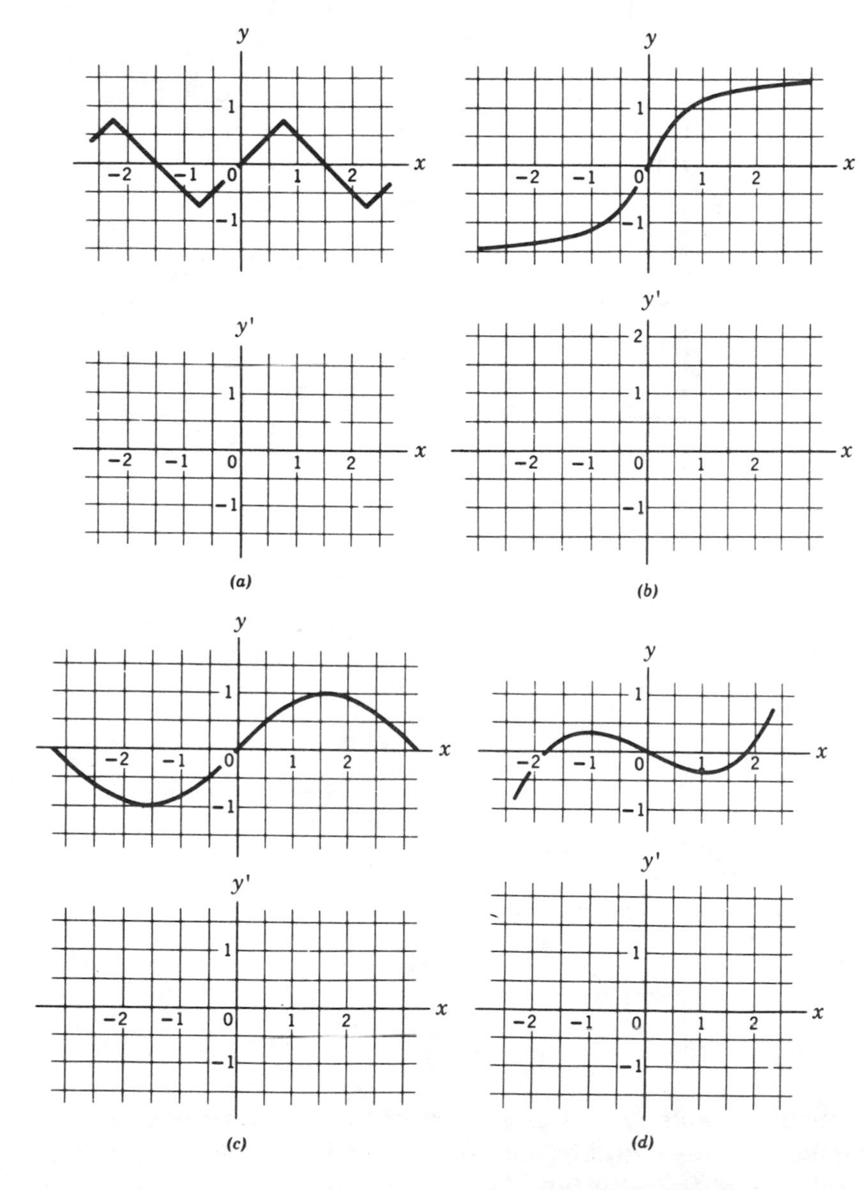

For the correct sketches, go to **169**.

169

Here are the solutions to the problems in frame **168**.

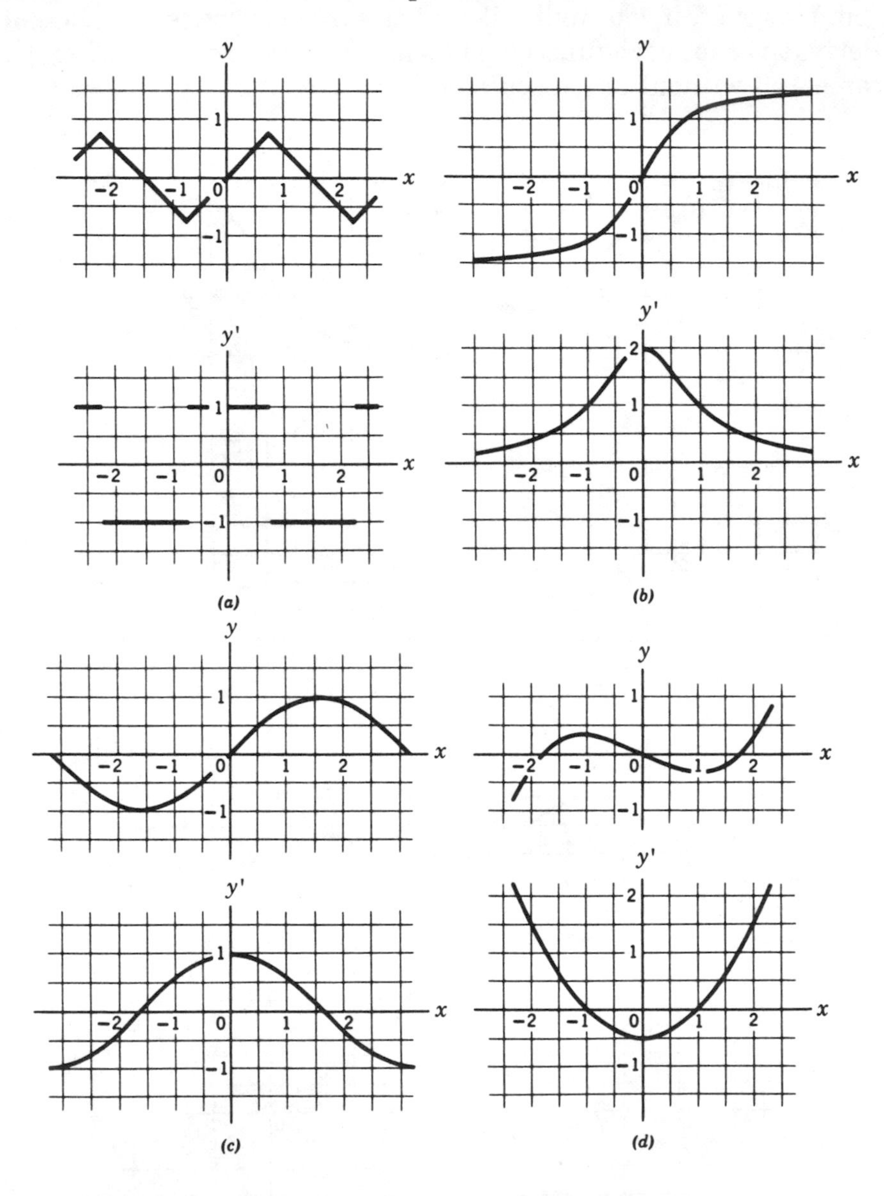

You should be able to convince yourself that the curves for y' have the general features we expect by comparing y' with the slope of a tangent to the graph of $y = f(x)$ at a few particular values of x.

Go to **170**.

Differentiation

170

We have accomplished a great deal so far in this chapter. In fact, all the really important new ideas involved in differential calculus have been introduced—limits, slopes of curves, and derivatives—and you are equipped in principle to solve a wide variety of problems. However, using the fundamental definition to calculate the derivative in each problem as it comes along would be very time-consuming. It would also be a great waste of time since there are numerous rules and tricks for differentiating apparently complicated functions in a few short steps. You will learn the most important of these rules in the following sections. You will also learn how to differentiate a few functions which occur so often that it is useful to know and remember their derivatives. These include a few of the trigonometric functions, logarithms, and exponentials. The remaining sections cover some special topics, as well as applications of differential calculus to some problems. By the end of this chapter you should be able to use differential calculus for many applications. Well, let's get going!

On to **171**.

171

Can you find the derivative of the following simple function?

$$y = a \qquad (a \text{ is a constant}).$$

$$y' = [1 \mid x \mid a \mid 0 \mid \text{none of these}]$$

If right, go to **173**.
If wrong, go to **172**.

172

To find y', we go back to the definition $\frac{dy}{dx} = \lim_{\Delta x \to 0} \frac{\Delta y}{\Delta x}$. If $y = a$,

$$\frac{\Delta y}{\Delta x} = \frac{f(x + \Delta x) - f(x)}{\Delta x} = \frac{a - a}{\Delta x} = 0.$$

(Remember that the meaning of $f(x + \Delta x)$ is f evaluated at $x + \Delta x$.)

(continued)

$$\lim_{\Delta x \to 0} \frac{\Delta y}{\Delta x} = \lim_{\Delta x \to 0} 0 = 0.$$

Since $y' = 0$, the plot of y in terms of x has 0 *slope*. (Figure 4 in frame **32** shows this graphically.)

Go to **173**.

173

You have just seen that the derivative of a constant is 0. Now, try to find the derivative of this function:

$$y = ax \qquad (a = \text{constant}).$$

$$\frac{dy}{dx} = [1 \mid x \mid a \mid 0 \mid ax \mid \text{none of these}]$$

If right, skip to **175**.
If wrong, go to **174**.

174

Here is the correct procedure:

$$y(x) = ax,$$

$$y(x + \Delta x) = a(x + \Delta x) - x = (ax + a\,\Delta x) - ax = a\,\Delta x.$$

Therefore

$$\frac{dy}{dx} = \lim_{\Delta x \to 0} \frac{\Delta y}{\Delta x} = \lim_{\Delta x \to 0} \frac{a\,\Delta x}{\Delta x} = a.$$

Now try to find the derivative of the function $f = -x$.

$$f' = [1 \mid 0 \mid a \mid -1 \mid -x]$$

If correct, go to **175**. If wrong, note that this problem is just a special case of **173**. Try again and then

Go to **175**.

Answer: (171) 0

175

Now we are going to find the derivative of a quadratic function. Suppose

$$y = f(x) = x^2.$$

What is y'?

You should be able to work this out from the definition of the derivative. Choose the correct answer:

$$y' = [1 \mid x \mid 0 \mid x^2 \mid 2x]$$

If right, go to **177**.
Otherwise, go to **176**.

176

Let us recall the definition of the derivative

$$y' = \frac{dy}{dx} = \lim_{\Delta x \to 0} \frac{y(x + \Delta x) - y(x)}{\Delta x}.$$

In this case, $y(x + \Delta x) = (x + \Delta x)^2 = x^2 + 2x\,\Delta x = (\Delta x)^2$, so

$$\lim_{\Delta x \to 0} \frac{y(x + \Delta x) - y(x)}{\Delta x} = \lim_{\Delta x \to 0} \frac{[x^2 + 2x\,\Delta x + (\Delta x)^2] - x^2}{x}$$

$$= \lim_{\Delta x \to 0} \frac{2x\,\Delta x + (\Delta x)^2}{\Delta x}$$

$$= \lim_{\Delta x \to 0} (2x + \Delta x) = 2x,$$

$$y' = \frac{dy}{dx} = 2x.$$

Go to **177**.

177

We have found the result that $\frac{d}{dx}(x^2) = 2x$. To illustrate this, a graph of $y = x^2$ is drawn in the figure. Since the slope of the curve at a point is simply the derivative at that point, each of the straight lines tangent to the curve has a slope equal to the derivative evaluated at the point of tangency.

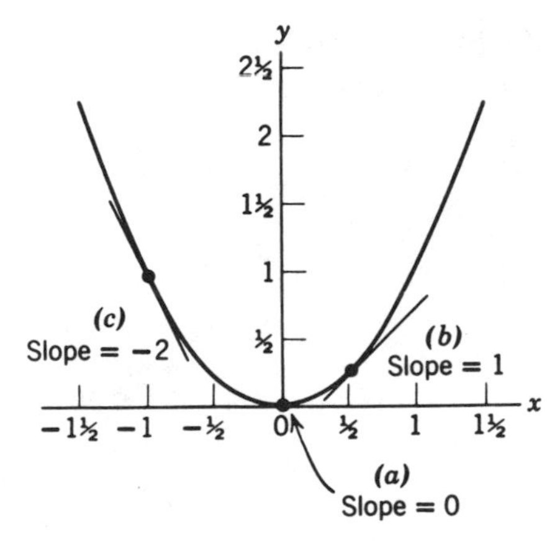

The tangent through the origin has a slope of (2)(0) = 0. Line (*b*) passes through the point $x = \frac{1}{2}$, and has slope $(2)(\frac{1}{2}) = 1$. Line (*c*) passes through the point $x = -1$, and has slope $(2)(-1) = -2$.

Go to **178**.

178

Here is a problem which summarizes the results we have had so far in this section (with a tiny bit of new material).

If $f = 3x^2 + 7x + 2$,

Find f'.

Answer: f' = ______________.

See frame **179** for the correct answer.

Answers: (173) *a* **(174)** − 1 **(175)** $2x$

179

If $f = 3x^2 + 7x + 2$, then $f' = 6x + 7$.

Congratulations if you got this answer. Go on to **180**. Otherwise, read below.

After you have finished this chapter, you will know several shortcuts for evaluating this derivative. However, right now we will use the basic definition:

$$f' = \frac{df}{dx} = \lim_{\Delta x \to 0} \frac{\Delta f}{\Delta x},$$

$$f(x) = 3x^2 + 7x + 2,$$

$$f(x + \Delta x) = 3[x^2 + 2x\Delta x + (\Delta x)^2] + 7(x + \Delta x) + 2,$$

$$\Delta f = f(x + \Delta x) - f(x) = 6x\,\Delta x + 3\,\Delta x^2 + 7\,\Delta x,$$

so

$$\frac{df}{dx} = \lim_{\Delta x \to 0} \left(\frac{6x\,\Delta x + 3\,\Delta x^2 + 7\,\Delta x}{\Delta x} \right) = \lim_{\Delta x \to 0} (6x + 3\,\Delta x + 7)$$

$$= 6x + 7.$$

Go to **180**.

180

Now that we have found the derivatives of x and x^2, our next step is to find the derivative of x^n, where n is any number. We will state the rule here, but you can look in Appendix A4 if you would like to see how it is derived.

The result is

$$\boxed{\frac{dx^n}{dx} = nx^{n-1}.}$$

This important result holds for all values of n: positive, negative, integral, fractional, irrational, etc. Note that our previous result, $\frac{d}{dx}(x^2) = 2x$, is the particular case of this when $n = 2$.

[Also, $\frac{d}{dx}(x) = 1$ is the particular case when $n = 1$.]

Go to **181**.

181

Now for a few applications.

Find $\frac{dy}{dx}$ for each of the following functions.

$$y = x^3, \qquad \frac{dy}{dx} = [3x^3 \mid 3x^2 \mid 2x^3 \mid x^2]$$

$$y = x^{-7}, \qquad \frac{dy}{dx} = [-7x^{-6} \mid 7x^{-7} \mid -7x^{-8} \mid -6x^{-7}]$$

$$y = \frac{1}{x^2}, \qquad \frac{dy}{dx} = [-2x \mid \frac{2}{x} \mid -\frac{2}{x^3}]$$

If all these were correct, go to **183**.
If you made any errors, go to **182**.

182

The solutions to these problems depend directly on the rule in frame **180**. Here are the details.

We use our general rule: $\frac{d}{dx}(x^n) = nx^{n-1}$.

$y = x^3$; in this case $n = 3$, so

$$\frac{d}{dx}(x^3) = 3x^{3-1} = 3x^2.$$

$y = x^{-7}$; here $n = -7$, so

$$\frac{d}{dx}(x^{-7}) = -7x^{-7-1} = -7x^{-8}.$$

$y = 1/x^2 = x^{-2}$; here $n = -2$, so

$$\frac{d}{dx}\left(\frac{1}{x^2}\right) = -2x^{-2-1} = -2x^{-3} = \frac{-2}{x^3}.$$

Now try these problems:

$$y = \frac{1}{x}, \qquad \frac{dy}{dx} = [1 + \frac{1}{x} \mid -\frac{1}{x} \mid -\frac{1}{x^2} \mid 2]$$

$$y = \frac{-1}{3}x^{-3}, \qquad \frac{dy}{dx} = [x^{-4} \mid -3x^{-4} \mid \frac{-1}{4}x^{-2} \mid +x^{-2}]$$

If right, go on to **183**.
If wrong, go back to **180** and continue from there.

183

Here is another application.

If $y = x^{1/2}$, find $\frac{dy}{dx}$.

The answer is. [$x^{-1/2}$ | $\frac{1}{2}x^{-1/2}$ | $\frac{1}{2}x$ | none of these].

If right, go to **185**.
If wrong, go to **184**.

184

The rule $\frac{dx^n}{dx} = nx^{n-1}$ is true for any value of n.

In this case, $n = \frac{1}{2}$,

$$\frac{d}{dx}(x^{1/2}) = \frac{1}{2}x^{(1/2-1)} = \frac{1}{2}x^{-1/2}.$$

Try this problem:

$$\frac{d}{dx}(x^{2/3}) = [x^{-1/3} \mid \frac{2}{3}x^{-2/3} \mid \frac{2}{3}x^{-1/3} \mid x^{5/3}]$$

Go to **185**.

Some Rules for Differentiation

185

In this section we are going to learn a number of shortcut rules for differentiation without having to go all the way back to the definition of the derivative each time. Some of these rules are derived here, while others are derived in Appendix A.

For the rest of this section, we will let $u(x)$ and $v(x)$ stand for any two variables that depend on x.

Go to **186**.

186

Our first rule will let us evaluate the derivative of the sum of u and v, in terms of their derivatives. We will derive the rule here. Let

$$y = u(x) + v(x).$$

Then

$$\begin{aligned}\frac{dy}{dx} &= \lim_{\Delta x \to 0} [u(x + \Delta x) + v(x + \Delta x) - u(x) - v(x)] \frac{1}{\Delta x} \\ &= \lim_{\Delta x \to 0} [u(x + \Delta x) - u(x)] \frac{1}{\Delta x} + \lim_{\Delta x \to 0} [v(x + \Delta x) - v(x)] \frac{1}{\Delta x} \\ &= \frac{du}{dx} + \frac{dv}{dx}.\end{aligned}$$

Hence

$$\boxed{\frac{d}{dx}(u + v) = \frac{du}{dx} + \frac{dv}{dx}.}$$

If you would like a rigorous justification of the manipulation of the limits in the above proof, see Appendix A2.

Go to **187**.

187

Now let's put the above rule to use by computing the derivative of the following function (you will also have to use some results from the last section):

$$y = x^4 + 8x^3.$$

$$\frac{dy}{dx} = \underline{\hspace{6cm}}.$$

For the correct answer, go to frame **188**.

Answers: **(181)** $3x^2, -7x^{-8}, -2/x^3$ **(182)** $-1/x^2, x^{-4}$

(183) $\frac{1}{2}x^{-1/2}$ **(184)** $\frac{2}{3}x^{-1/3}$

188

The correct answer to the question in frame **187** is

$$\frac{d}{dx}(x^4 + 8x^3) = 4x^3 + 24x^2.$$

If you got this answer, go to frame **189**. Otherwise, continue here to find your mistake.

Our problem is to find the derivative of the sum of two functions. To make use of the rule in frame **186** in the notation used there, suppose we let $u = x^4$, $v = 8x^3$.

Then

$$\frac{d}{dx}(u + v) = \frac{d}{dx}(x^4 + 8x^3) = \frac{d}{dx}(x^4) + \frac{d}{dx}(8x^3).$$

You should be able to evaluate these two derivatives from the result of the last section:

$$\frac{d}{dx}(x^4) = 4x^3, \qquad \frac{d}{dx}(8x^3) = 24x^2.$$

Hence, $\frac{d}{dx}(x^4 + 8x^3) = 4x^3 + 24x^2$.

Go to **189**.

189

Now that we can differentiate the sum of two variables, our next task is to learn to differentiate the product, for instance, $u(x) \times v(x)$. We want to express $\frac{d}{dx}(uv)$ in terms of $\frac{du}{dx}$ and $\frac{dv}{dx}$. The result, known as the *product rule,* will be stated here. Look in Appendix A6 if you want to see how it is derived.

Product rule

$$\boxed{\frac{d}{dx}(uv) = u\frac{dv}{dx} + v\frac{du}{dx} = uv' + vu'.}$$

Go to **190**.

190

Here is an example in which the *product rule* is used. Suppose

$$y = (x^5 + 7)(x^3 + 17x).$$

The problem is to find $\frac{dy}{dx}$. If we let $u = x^5 + 7$, $v = x^3 + 17x$, then $y = uv$.

$$\frac{dy}{dx} = \frac{d}{dx}(uv) = u\frac{dv}{dx} + v\frac{du}{dx}.$$

Since $\frac{du}{dx} = 5x^4$ and $\frac{dv}{dx} = 3x^2 + 17$, our result is

$$\frac{dy}{dx} = (x^5 + 7)(3x^2 + 17) + (x^3 + 17x)(5x^4).$$

Note that it is usually considered good practice to simplify (collect together terms in like powers of x) expressions such as this. However, to save time in this chapter, you need not do so.

By using the product rule, we can derive in another way a result we have already found: $\frac{d}{dx}(x^2) = 2x$. If we let $u = x$ and $v = x$, then the product rule tells us that

$$\frac{d}{dx}(x^2) = x\frac{dx}{dx} + x\frac{dx}{dx} = 2x.$$

Go to **191**.

191

Use the product rule to find the derivative $\frac{d}{dx}[(3x + 7)(4x^2 + 6x)]$.

Answer: ____________________

Go to **192** for the solution.

192

The answer is

$$(3x + 7)(8x + 6) + (4x^2 + 6x)(3).$$

If you obtained this or an equivalent result, go on to **194**. Otherwise, read below.

The problem is to differentiate the product of $3x + 7$ and $4x^2 + 6x$. Suppose we let $u = 3x + 7$ and $v = 4x^2 + 6x$. Then $u' = 3$, $v' = 8x + 6$. Hence

$$\frac{d}{dx}(uv) = uv' + vu' = (3x + 7)(8x + 6) + (4x^2 + 6x)(3).$$

Try this problem:

What is $\frac{d}{dx}[(2x + 3)(x^5)]$?

Answer: ______________________________

Go to **193** for the correct solution.

193

$$\frac{d}{dx}[(2x + 3)(x^5)] = (2x + 3)(5x^4) + (x^5)(2).$$

The method for obtaining this is like that shown in frame **192**. You can use the rule in frame **180** for differentiating x^n in order to find $\frac{d}{dx}(x^5) = 5x^4$.

Go to **194**.

194

In frame **189** we learned the product rule: $(uv)' = uv' + vu'$. Sometimes one needs to differentiate the *quotient* of two functions, $u(x)/v(x)$. Here is the rule. It will be proven later in this section, in frame **206**.

Quotient rule

$$\frac{d}{dx}\left(\frac{u}{v}\right) = \frac{v(du/dx) - u(dv/dx)}{v^2} = \frac{vu' - uv'}{v^2}.$$

Go to **195**.

195

Solve the following problem:

$$\frac{d}{dx}\left(\frac{1+x}{x^2}\right) = \_\_\_\_\_\_\_\_\_\_$$

To see the correct answer, go to **196**.

196

The answer to the problem in **195** is

$$\frac{d}{dx}\left(\frac{1+x}{x^2}\right) = -\frac{2}{x^3} - \frac{1}{x^2}.$$

If right, go to **198**.
If wrong, you should go to **197** for help.

197

Let $u = 1 + x$, $v = x^2$. Then $\frac{du}{dx} = 1$, $\frac{dv}{dx} = 2x$.

$$\frac{d}{dx}\left(\frac{u}{v}\right) = \frac{v(du/dx) - u(dv/dx)}{v^2},$$

$$\frac{d}{dx}\left(\frac{u}{v}\right) = \frac{x^2 - (1 + x)(2x)}{x^4} = \frac{1}{x^2} - \frac{2}{x^3}(1 + x)$$

$$= -\frac{2}{x^3} - \frac{1}{x^2}.$$

Go to **198**.

198

In this frame we are going to learn a useful rule for finding the derivative of a "function of a function." Suppose f is a variable that depends on u, and u in turn depends on x. Then f also depends on x. The following rule is proved in Appendix A7.

Chain rule

$$\boxed{\frac{df}{dx} = \frac{df}{du}\frac{du}{dx}.}$$

This formula is called the *chain rule* because it links together derivatives with related variables. It is one of the most frequently used rules in differential calculus.

Here is an example: Suppose we want to differentiate $f(x) = (x + x^2)^2$. This is a complicated function. It looks much simpler if we let $u = x + x^2$, in which case $f(u) = u^2$.

$$\frac{df}{dx} = \frac{df}{du}\frac{du}{dx} = \frac{d}{du}(u^2)\frac{du}{dx} = 2u\frac{du}{dx}.$$

We now substitute the value $u = x + x^2$, and $\frac{du}{dx} = 1 + 2x$, to obtain

$$\frac{df}{dx} = 2(x + x^2)(1 + 2x).$$

(You can check that the chain rule gives the right answer in this case by multiplying out the expression for f, and then differentiating it. You will find that the answer is equivalent to $\frac{df}{dx}$ found above.)

(continued)

Caution: The chain rule would be a simple identity if $\frac{df}{dx}$ and $\frac{du}{dx}$ could be treated as ratios of independent quantities df, du, dx. However, this is not the case; one cannot cancel du's in the numerator and denominator. (Nevertheless, this fiction makes a very handy way to *remember* the chain rule!)

Go to **199**.

199

Here are a few more examples of the use of the *chain rule.*

1. Find $\frac{d}{dt}(\sqrt{1+t^2})$.

Suppose we let $w = \sqrt{1 + t^2}$, and $u = 1 + t^2$, so that $w = \sqrt{u}$. Then

$$\frac{dw}{dt} = \frac{dw}{du}\frac{du}{dt} = \frac{1}{2\sqrt{u}}(2t)$$

$$= \frac{1}{2}\frac{1}{\sqrt{1+t^2}}2t = \frac{t}{\sqrt{1+t^2}}.$$

Here we have used t as a variable, but of course it makes no difference what we call the variables.

2. Let $v = \left(q^3 + \frac{1}{q}\right)^{-3}$; find $\frac{dv}{dq}$.

This problem can be simplified by letting $p = q^3 + 1/q$ and $v = p^{-3}$. With these symbols the chain rule is

$$\frac{dv}{dq} = \frac{dv}{dp}\frac{dp}{dq} = -3p^{-4}\frac{dp}{dq} = -3p^{-4}\left(3q^2 - \frac{1}{q^2}\right)$$

$$= -3\left(q^3 + \frac{1}{q}\right)^{-4}\left(3q^2 - \frac{1}{q^2}\right).$$

The following example will not be explained, since you should be able to work it by inspection.

3. $\frac{d}{dx}\left(1 + \frac{1}{x}\right)^2 = 2\left(1 + \frac{1}{x}\right)\left(-\frac{1}{x^2}\right)$

Go to **200**.

200

Now try the following problem:

Which expression correctly gives

$$\frac{d}{dx}(2x + 7x^2)^{-2}?$$

(a) $-2(2 + 14x)^{-3}$
(b) $-2(2 + 14x)^{-2}(2x + 7x^2)$
(c) $(2x + 7x^2)^{-3}(2 + 14x)$
(d) $-2(2x + 7x^2)^{-3}(2 + 14x)$

The correct answer is [*a* | *b* | *c* | *d*]

If right, go to **203**.
Otherwise, go to **201**.

201

Here is how to work the problem in **200**. Suppose we let $w = u^{-2}$ and $u = 2x + 7x^2$. Then

$$\frac{du}{dx} = 2 + 14x.$$

Hence

$$\frac{dw}{dx} = \frac{dw}{du}\frac{du}{dx} = \frac{d}{du}(u^{-2})\frac{du}{dx}$$

$$= -2u^{-3}\frac{du}{dx} = -2(2x + 7x^2)^{-3}(2 + 14x).$$

Try this problem:

Find $\frac{dw}{ds}$, where $w = 12q^4 + 7q$, and $q = s^2 + 4$.

$$\frac{dw}{ds} = \underline{\hspace{4cm}}.$$

For the solution, go to **202**.

202

The problem in frame **201** can be solved by using the chain rule:

$$\frac{dw}{ds} = \frac{dw}{dq}\frac{dq}{ds}.$$

We are given that $w = 12q^4 + 7q$ and $q = s^2 + 4$, so

$$\frac{dw}{dq} = 48q^3 + 7 \qquad \text{and} \qquad \frac{dq}{ds} = 2s.$$

Substituting these, we have

$$\frac{dw}{ds} = (48q^3 + 7)(2s) = [48(s^2 + 4)^3 + 7](2s).$$

If you wrote this result, go on to **203**. If you made a mistake, you should study the last few frames to make sure you understand the application of the chain rule. Don't be confused by the names of variables.

Then go to **203**.

203

The next problem is to find $\frac{d}{dx}\left(\frac{1}{v}\right)$ in terms of v and $\frac{dv}{dx}$, where v depends on x. The answer can be found using the quotient rule, but since we are going to use it to *prove* the quotient rule, don't use that rule here. Instead, try the chain rule.

Which of the following answers correctly gives $\frac{d}{dx}\left(\frac{1}{v}\right)$?

[$-\frac{1}{v^2}\frac{dv}{dx}$ | $\frac{1}{dv/dx}$ | $\frac{dx}{dv}$ | $-\frac{dv}{dx}$ | none of these]

If right, go to **205**.
If wrong, go to **204**.

204

To find $\frac{d}{dx}\left(\frac{1}{v}\right)$, we apply the chain rule in the following way. Suppose we let $f = \frac{1}{v} = v^{-1}$. $\frac{df}{dx} = \frac{df}{dv}\frac{dv}{dx}$, but $\frac{df}{dv} = \frac{d}{dv}v^{-1} = -\frac{1}{v^2}$, so $\frac{d}{dx}\left(\frac{1}{v}\right) = -\frac{1}{v^2}\frac{dv}{dx}$.

Go to **205**.

Answer: (200) d

205

Now, by combining the result of the last frame with what you have learned previously, you should be able to derive the expression for the derivative of the quotient of two functions. This is an extremely important relation. Try to work it out for yourself.

Find $\frac{d}{dx}\left(\frac{u}{v}\right)$ in terms of u, v, $\frac{du}{dx}$, $\frac{dv}{dx}$.

$$\frac{d}{dx}\left(\frac{u}{v}\right) = \underline{\hspace{4cm}}.$$

To check your answer, go to **206**.

206

You should have obtained the following quotient rule which was presented without proof in frame **194**, though possibly arranged differently.

$$\frac{d}{dx}\left(\frac{u}{v}\right) = \frac{uv' - vu'}{v^2}.$$

If you wrote this or an equivalent statement, go on to **207**. Otherwise, study the derivation below.

If we let $p = \frac{1}{v}$, then our derivative is that of the product of two variables.

$$\frac{d}{dx}\left(\frac{u}{v}\right) = \frac{d}{dx}(up) = u\frac{dp}{dx} + p\frac{du}{dx}.$$

Now $\frac{dp}{dx} = \frac{dp}{dv}\frac{dv}{dx} = -\frac{1}{v^2}\frac{dv}{dx}$, as in frame **194**, so

$$\frac{d}{dx}\left(\frac{u}{v}\right) = -\frac{u}{v^2}\frac{dv}{dx} + \frac{1}{v}\frac{du}{dx} = \frac{v(du/dx) - u(dv/dx)}{v^2}.$$

Go to **207**.

207

Before going on to new material, let's summarize all the rules for differentiation we have used so far. Fill in the blanks. a and n are constants, u and v are variables that depend on x, w depends on u, which in turn depends on x.

$$\frac{d}{dx}(a) = \_\_\_\_\_\_\_\_\_\_\_\_. \qquad \frac{d}{dx}(u+v) = \_\_\_\_\_\_\_\_\_\_\_\_.$$

$$\frac{d}{dx}(ax) = \_\_\_\_\_\_\_\_\_\_\_\_. \qquad \frac{d}{dx}(uv) = \_\_\_\_\_\_\_\_\_\_\_\_.$$

$$\frac{d}{dx}(x^2) = \_\_\_\_\_\_\_\_\_\_\_\_. \qquad \frac{d}{dx}\left(\frac{u}{v}\right) = \_\_\_\_\_\_\_\_\_\_\_\_.$$

$$\frac{d}{dx}(x^n) = \_\_\_\_\_\_\_\_\_\_\_\_. \qquad \frac{d}{dx}[w(u)] = \_\_\_\_\_\_\_\_\_\_\_\_.$$

Go to **208**.

208

Here are the correct answers. The frame in which the relation was introduced is shown in parentheses.

$$\frac{d}{dx}(a) = 0. \quad \textbf{(172)} \qquad \frac{d}{dx}(u+v) = \frac{du}{dx} + \frac{dv}{dx}. \quad \textbf{(186)}$$

$$\frac{d}{dx}(ax) = a. \quad \textbf{(174)} \qquad \frac{d}{dx}\left(\frac{u}{v}\right) = u\frac{dv}{dx} + v\frac{du}{dx} \quad \textbf{(189)}$$

$$\frac{d}{dx}(x^2) = 2x. \quad \textbf{(176)} \qquad \frac{d}{dx}\left(\frac{u}{v}\right) = \frac{v(du/dx) - u(dv/dx)}{v^2}. \quad \textbf{(194)}$$

$$\frac{d}{dx}(x^n) = nx^{n-1}. \quad \textbf{(180)} \qquad \frac{d}{dx}[w(u)] = \frac{dw}{du}\frac{du}{dx}. \quad \textbf{(198)}$$

If you would like some more practice on problems similar to those in the last two sections, see review problems 34 through 38.

Go to **209**.

Answer: (203) $-\frac{1}{v^2}\frac{dv}{dx}$

Differentiating Trigonometric Functions

209

Trigonometric functions occur in so many applications that it is useful to know their derivatives. For instance, we would like to know $\frac{d}{d\theta}(\sin \theta)$. By definition,

$$\frac{d}{d\theta}(\sin \theta) = \lim_{\Delta\theta \to 0} \frac{\sin(\theta + \Delta\theta) - \sin \theta}{\Delta\theta}.$$

It is not at all obvious how to evaluate this expression, so let's take another approach for a minute and try to guess *geometrically* what the result should be by looking at a plot of sin θ.

Here is a plot of sin θ vs. θ over the interval $0 \leq \theta \leq 2\pi$. (θ is measured in radians, but for reference, a few of the angles are shown in degrees.)

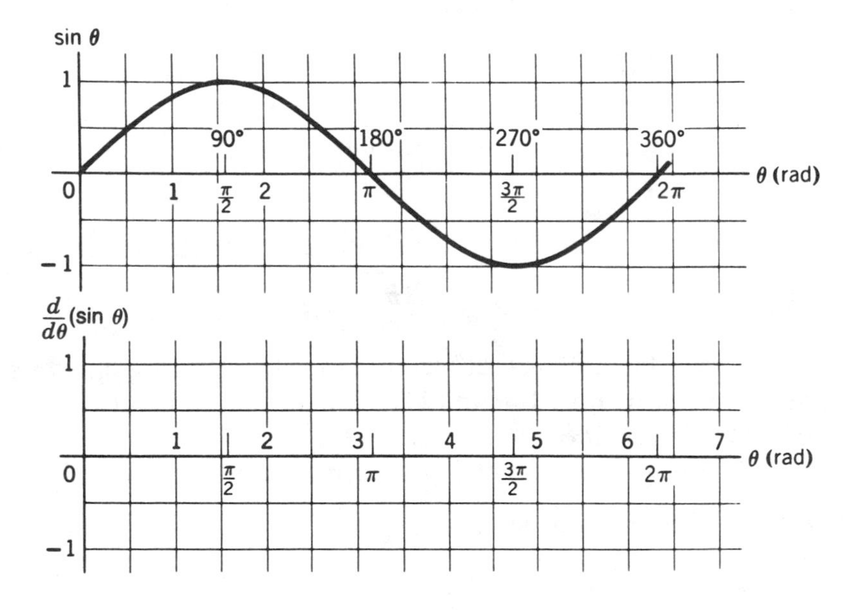

Draw a sketch of $\frac{d}{d\theta}(\sin \theta)$ in the space provided. To check your sketch,

Go to **210**.

210

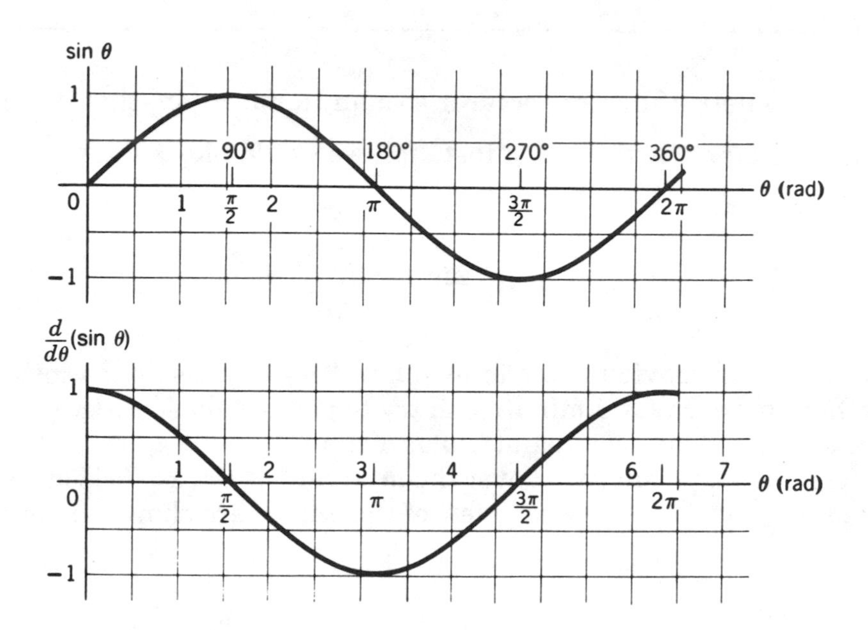

Here are drawings of $\sin\theta$ and $\frac{d}{d\theta}(\sin\theta)$. Note that where the slope of $\sin\theta$ is greatest, at 0 and 2π, $\frac{d}{d\theta}(\sin\theta)$ has its greatest value, and that where the slope is 0, at $\theta = \frac{\pi}{2}$ and $\frac{3\pi}{2}$, $\frac{d}{d\theta}(\sin\theta)$ is 0.

[If your sketch looked very different from the drawing shown above, you should review frames **160** and **169**. This problem is quite similar to problem (c) in frame **168**.]

Now, by looking at the graphs, you may be able to guess the correct answer for $\frac{d}{d\theta}(\sin\theta)$. Can you?

$$\frac{d}{d\theta}(\sin\theta) = \_\_\_\_\_\_\_\_\_\_\_\_.$$

Go to frame **211** to see if your answer is right.

211

Here is the rule:

$$\frac{d}{d\theta}(\sin\theta) = \cos\theta.$$

Congratulations if you guessed this result in the last frame! If you arrived at some other result, study the drawings in frame **209** and compare the second one with the graph of $\cos\theta$ shown below.

Formal proof that $\frac{d}{d\theta}(\sin\theta) = \cos\theta$ is given in Appendix A4. It is important to realize that this relation is only true when angle is measured in radians—this is why the radian is such a useful unit.

Let's try to guess the result for $\frac{d}{d\theta}(\cos\theta)$ from a plot of $\cos\theta$.

Draw a sketch of $\frac{d}{d\theta}(\cos\theta)$ in the space provided, and make a guess at the result.

$$\frac{d}{d\theta}\cos\theta = \_\_\_\_\_\_\_\_\_\_.$$

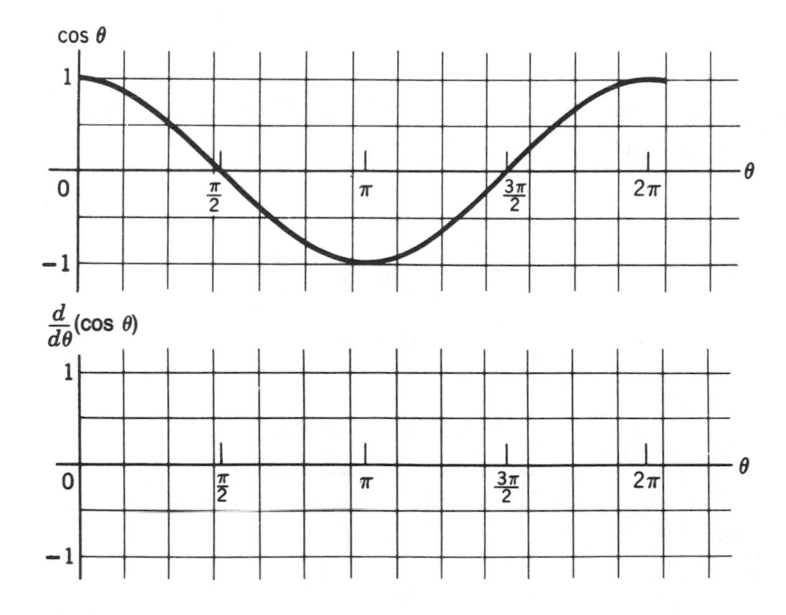

Go to **212**.

212

Here are plots of $\cos\theta$ and $\frac{d}{d\theta}(\cos\theta)$. The result is $\frac{d}{d\theta}(\cos\theta) = -\sin\theta$,

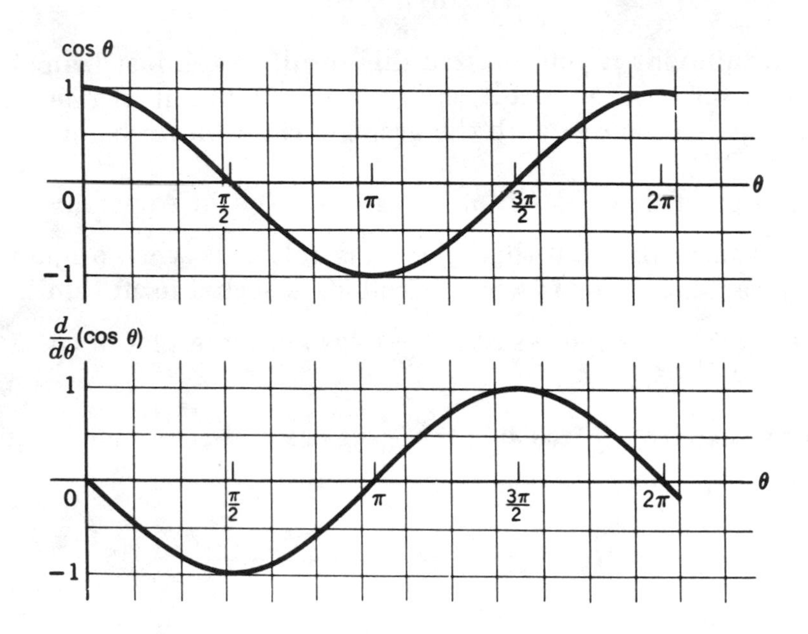

as should seem reasonable from the graph. This relation also is formally proved in Appendix A5.

To summarize:

$$\frac{d}{d\theta}(\sin\theta) = \cos\theta.$$

$$\frac{d}{d\theta}(\cos\theta) = -\sin\theta.$$

Using these results, find $\frac{d}{d\theta}(\tan\theta)$. (*Hint:* use $\tan\theta = \frac{\sin\theta}{\cos\theta}$ and apply the quotient rule, frame **194**.)

$$\frac{d}{d\theta}(\tan\theta) = \rule{4cm}{0.4pt}.$$

Go to **213**.

213

Using the hints in frame **212** we have

$$\frac{d}{d\theta}(\tan\theta) = \frac{d}{d\theta}\left(\frac{\sin\theta}{\cos\theta}\right)$$

$$= \frac{\cos\theta\,\frac{d}{d\theta}(\sin\theta) - \sin\theta\frac{d}{d\theta}(\cos\theta)}{\cos^2\theta}$$

$$= \frac{\cos^2\theta + \sin^2\theta}{\cos^2\theta} = \frac{1}{\cos^2\theta} = \sec^2\theta.$$

Now find the correct answer:

$$\frac{d}{d\theta}(\sec\theta) = [\sec\theta\tan\theta \mid -\sec\theta\tan\theta \mid \sec\theta]$$

If right, go to **215**.
If wrong, go to **214**.

214

Using the definition $\sec\theta = \frac{1}{\cos\theta}$, and the result in frame **204**, we have

$$\frac{d}{d\theta}(\sec\theta) = \frac{d}{d\theta}\left(\frac{1}{\cos\theta}\right) = -\frac{1}{\cos^2\theta}\frac{d\cos\theta}{d\theta}$$

$$= +\frac{1}{\cos^2\theta}\sin\theta = \frac{\tan\theta}{\cos\theta}$$

$$= \sec\theta\tan\theta.$$

(All three of these expressions are equally acceptable.)

Go to **215**.

215

Choose the correct answer:

$$\frac{d}{d\theta}(\sin\theta)^2 = [\sin\theta \mid 2\cos\theta \mid \cos\theta^2 \mid 2\sin\theta\cos\theta]$$

If right, go to **217**.
If wrong, go to **216**.

216

You could have analyzed the problem as follows:

Suppose we let $u(\theta) = \sin\theta$. Then $\frac{du}{d\theta} = \cos\theta$, and

$$\frac{d}{d\theta}(\sin\theta)^2 = \frac{d}{d\theta}u^2 = \frac{d}{du}(u^2)\frac{du}{d\theta}$$

$$= 2u\frac{du}{d\theta} = 2\sin\theta\cos\theta.$$

Where did you go wrong? Find your error and be sure you understand it. Then

Go to **217**.

Answers: (213) $\sec\theta\tan\theta$

217

Which of the following is $\frac{d}{d\theta}(\cos\theta^3)$?

[$\cos\theta\sin\theta^3$ | $-3\theta^2\sin\theta^3$ | $3\cos^2\theta^3\sin\theta^3$ | $3\cos^2\theta$]

If right, skip on to frame **221**.
If wrong, go to frame **218**.

218

Did you forget how to use the *chain rule* to differentiate a function of a function? We can think of $\cos\theta^3$ as a function of a function. Suppose we write it this way:

$$w = \cos u, \qquad u = \theta^3.$$

Then

$$\frac{dw}{d\theta} = \frac{dw}{du}\frac{du}{d\theta},$$

$$\frac{dw}{du} = -\sin u = -\sin\theta^3, \qquad \frac{du}{d\theta} = 3\theta^2,$$

so

$$\frac{d}{d\theta}(\cos\theta^3) = -3\theta^2\sin\theta^3.$$

Go to **219**.

219

If ω (Greek letter omega) is a constant, which expression correctly gives $\frac{d}{dt}(\sin\omega t)$?

[$\cos\omega t$ | $\omega\cos\omega t$ | $\sin\omega t$ | none of these]

If right, go to frame **221**.
Otherwise, go to **220**.

220

To solve problem in **219**, let $w = \sin u$, $u = \omega t$,

$$\frac{dw}{dt} = \frac{dw}{du}\frac{du}{dt} = \cos u \times \frac{d}{dt}(\omega t) = \omega \cos \omega t.$$

Go to frame **221**.

221

Before you go on to the next section, let's state once more the important relations we have introduced in this section:

$$\frac{d}{d\theta}(\sin \theta) = \cos \theta,$$
$$\frac{d}{d\theta}(\cos \theta) = -\sin \theta.$$

There are two more functions which are so common that it is worth knowing their derivatives by heart: logarithmic and exponential. To learn about them,

Go to **222**.

Differentiation of Logarithms and Exponentials

222

Our next task is to learn how to differentiate logarithms. If you feel shaky about logarithms, review frames **75–95** of Chapter 1 before going on to the next frame.

Go to **223**.

Answers: (**214**) $2 \sin \theta \cos \theta$
(**217**) $-3\theta^2 \sin \theta^3$
(**219**) $\omega \cos \omega t$

223

In this section we are going to work with natural logarithms, $\ln x = \log_e x$. Natural logarithms were defined in frame **94**. The base $e = 2.71828\ldots$ was discussed in frame **109**.

Here is a table showing $\ln x$ for a few values of x.

x	$\ln x$	x	$\ln x$
1	0.000	30	3.40
2	0.69	100	4.61
e	1.00	300	5.70
3	1.10	1000	6.91
10	2.30	3000	8.01

Using the table and the rules for manipulating logarithms, find the answer which is most nearly correct for each of the following questions:

$$\ln 6 = [2.2 \mid 3.1 \mid 6/e \mid 1.79]$$

$$\ln \sqrt{10} = [1.15 \mid 2.35 \mid 2.25 \mid 1.10]$$

$$\ln 300^3 = [126 \mid 185 \mid 17.10 \mid 3.41]$$

If all your answers are correct, go to **225**.
If you made any mistakes, go to **224**.

224

The rules for manipulating logarithms are summarized in frame **91**. These rules apply to logarithms of all bases, including the base e.

$$\ln 6 = \ln (2 \times 3) = \ln 2 + \ln 3 = 0.69 + 1.10 = 1.79,$$

$$\ln \sqrt{10} = \ln 10^{1/2} = \tfrac{1}{2} \ln 10 = \tfrac{1}{2} \times 2.30 = 1.15,$$

$$\ln 300^3 = 3 \ln 300 = 3 \times 5.70 = 17.10.$$

Go on to **225**.

225

Here is a plot of $\ln x$ in terms of x. If your calculator provides $\ln x$, check some of the points on this graph.

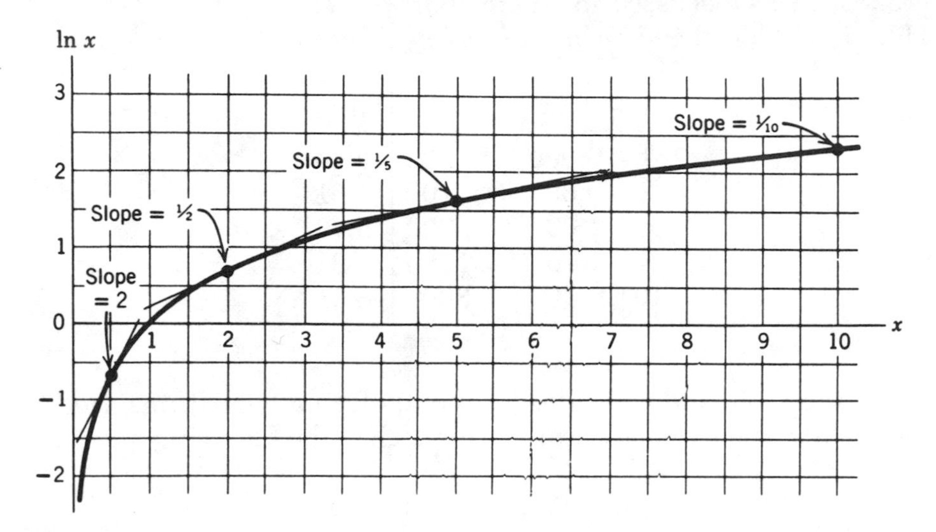

You can find the qualitative features of $\frac{d}{dx}(\ln x)$ by looking at the graph. For small values of x the derivative is large, and for large values of x the derivative is small. In the figure above tangents are shown at a few points, and their slopes are listed in this table.

x	Slope
½	2
2	½
5	⅕
10	⅒

Perhaps you can guess the formula for $\frac{d}{dx}(\ln x)$. Try to fill in the blank.

$$\frac{d}{dx}(\ln x) = \_\_\_\_\_\_\_\_\_\_.$$

To learn the correct expression, go to **226**.

Answers: (223) 1.79, 1.15, 17.10

226

Here is the formula for the derivative of a natural logarithm:

$$\boxed{\frac{d}{dx}(\ln x) = \frac{1}{x}.}$$

If you did not guess this result, you can check that it agrees with the numerical values in the table in frame **225**.

The reason that e is so useful as a base for logarithms is that it leads to such a simple expression. This relation is derived in Appendix A9. It is so important that it is worth committing to memory.

Go to **227**.

227

Skip on to frame **228** if you do not have a scientific calculator. Using a calculator, you can numerically confirm that $\frac{d}{dx}(\ln x) = \frac{1}{x}$. The procedure is to calculate value of $[\ln (x + \Delta) - \ln x]/\Delta$ for successively smaller values of Δ. The result should approach $1/x$.

Try the following for $x = 5$, for instance, or any other value you may wish to choose. For $x = 5$, $\ln x = 1.6094$ and $\ln' x = 1/5 = 0.2$.

Δ	$\ln(x + \Delta)$	$\frac{\ln(x + \Delta) - \ln x}{\Delta}$
2		
1		
0.1		
0.01		

Go to **228**.

228

Try this problem: Which of the following gives $\frac{d}{dx}(\ln x^2)$?

$$\left[2 \ln x \;\middle|\; \frac{2}{x} \;\middle|\; \frac{1}{x^2} \;\middle|\; \frac{2}{x^2} \;\middle|\; \frac{2}{x} \ln x\right]$$

If right, go to **230**.
Otherwise, go to **229**.

229

The solution of this problem is quite straightforward. We could make use of the chain rule. However, let's solve it another way.

Since $\ln x^2 = 2 \ln x$,

$$\frac{d}{dx}(\ln x^2) = \frac{d}{dx} 2 \ln x = \frac{2}{x}.$$

You should be able to do this one:

$$\frac{d}{dx}(\ln x)^2 = \left[2 \ln x \;\middle|\; \frac{2 \ln x}{x} \;\middle|\; \frac{2}{\ln x} \;\middle|\; \text{none of these}\right]$$

If right, go to **231**.
Otherwise, go to **230**.

230

$$\frac{d}{dx}(\ln x)^2 = 2 \ln x \frac{d}{dx}(\ln x) = \frac{2 \ln x}{x}.$$

Go to **231**.

231

(a) $\dfrac{d \ln r}{dr} =$ ________________.

(b) $\dfrac{d \ln 5z}{dz} =$ ________________.

For the correct answers, go to **232**.

232

The correct answers are

$$\text{(a) } \frac{1}{r}; \qquad \text{(b) } \frac{1}{z}.$$

If you got both of these, you are doing fine, so you may skip ahead to frame **234**. If you missed either one,

Go to frame **233**.

233

(a) $\dfrac{d \ln r}{dr} = \dfrac{1}{r}$ for the same reason that $\dfrac{d \ln x}{dx} = \dfrac{1}{x}$. It makes no difference whether the variable is called r or x.

(b) The simplest way to find $\dfrac{d}{dx}(\ln 5z)$ is to recall that $\ln 5z = \ln 5 + \ln z$. Hence,

$$\frac{d}{dz}(\ln 5z) = \frac{d}{dz}(\ln 5) + \frac{d}{dz}(\ln z) = 0 + \frac{1}{z} = \frac{1}{z}.$$

Go to **234**.

234

Another function we would like to differentiate is

$$y = a^x \qquad (a \text{ is a constant}).$$

(*Warning:* Do not confuse a^x with x^a, where x is a variable and a is a constant.)

We can differentiate a^x by taking the natural logarithm:

$$\ln y = \ln a^x = x \ln a.$$

Now differentiate both sides of this equation with respect to x:

$$\frac{d}{dx}(\ln y) = \frac{dx}{dx} \ln a,$$

$$\frac{1}{y}\frac{dy}{dx} = \ln a,$$

$$\frac{dy}{dx} = y \ln a = a^x \ln a.$$

Thus,

$$\frac{d}{dx}(a^x) = a^x \ln a.$$

Go to **235**.

235

The preceding frame gave the result

$$\frac{da^x}{dx} = a^x \ln a$$

A particularly simple but important case occurs when $a = e$. Since $\ln e = 1$,

$$\boxed{\frac{de^x}{dx} = e^x.}$$

Answers: (228) $\frac{2}{x}$ **(230)** $\frac{2 \ln x}{x}$

With the above, can you write the values for the following?

(a) $\frac{de^{cx}}{dx} =$ ______________

(b) $\frac{de^{-x}}{dx} =$ ______________

See **236** for the correct answers.

236

The answers are

(a) $\frac{de^{cx}}{dx} = ce^{cx}$

and

(b) $\frac{de^{-x}}{dx} = -e^{-x}.$

If you did both of these correctly, go to **237**. Otherwise, continue here.
The result (a) is obtained by letting $u = cx$ and following the usual procedure for a function of a function (i.e., using the chain rule, frame **194**). Thus

$$\frac{de^{cx}}{dx} = \frac{de^{u}}{du}\frac{du}{dx} = e^{u}c = ce^{cx}.$$

The result (b) is a special case of (a) with $c = -1$.

Go to **237**.

237

Skip on to frame **238** if you do not have a scientific calculator.

You can confirm numerically that $\frac{d}{dx}(e^x) = e^x$ in the same way that you confirmed $\frac{d}{dx}(\ln x) = \frac{1}{x}$ in frame **227**. Calculate the following for some value of x, for instance, $x = 10$. See whether the last column approaches $e^{10} = 22{,}026.46\ldots$

Δ	$e^{x+\Delta}$	$\frac{e^{x+\Delta} - e^x}{\Delta}$
1		
0.1		
0.01		

Go to **238**.

238

If $z = \frac{1}{\ln x}$, what is $\frac{dz}{dx}$?

Encircle the correct answer.

$$\left[\frac{1}{x \ln x} \quad \Big| \quad \frac{-x}{(\ln x)^2} \quad \Big| \quad \frac{-1}{x(\ln x)^2} \quad \Big| \quad \frac{\ln x}{x^2}\right]$$

If right, go to **240**.
Otherwise, go to **239**.

239

One way to find the derivative of $\frac{1}{\ln x}$ is to use the chain rule. Let $u = \ln x$. Then

$$\frac{d}{dx}\left(\frac{1}{\ln x}\right) = \frac{d}{dx}\left(\frac{1}{u}\right) = \frac{du^{-1}}{du}\frac{du}{dx} = -\frac{1}{u^2}\frac{1}{x}$$

$$= -\frac{1}{x(\ln x)^2}.$$

Go to **240**.

240

A number of relations have been used in this section and you may want to give them a quick review before going on. Here is a list. The most important ones are in boxes.

$$e = 2.71828\ldots,$$

$$\ln x = \log_e x,$$

$$\ln(x) = 2.303\ldots \log_{10} x,$$

$$\boxed{\frac{d}{dx}(\ln x) = \frac{1}{x},}$$

$$\frac{d}{dx}(a^x) = a^x \ln a,$$

$$\boxed{\frac{d}{dx}(e^x) = e^x.}$$

Go to **241**.

241

We have learned how to differentiate the most useful common functions. The rest of this chapter will be spent on some special topics related to the use of derivatives. However, you may want a little more practice in differentiation before you go on. If so, see problems 34 through 58 on page 247. Whenever you are ready,

Go to **242**.

Higher-Order Derivatives

242

Suppose y depends on x and we have obtained the derivative $\frac{dy}{dx}$. If we next differentiate $\frac{dy}{dx}$ with respect to x, the result is called the *second derivative* of y with respect to x, and is written $\frac{d^2y}{dx^2}$.

Can you do the following problem?

$$\text{If } y = 2x^3, \text{ then } \frac{d^2y}{dx^2} = [6x^2 \mid 12x \mid 0 \mid x^2 \mid x]$$

If right, go to **245**.
If wrong, go to **243**.

243

Here's how to do the problem in **242**.

$$y = 2x^3,$$

$$\frac{dy}{dx} = 6x^2,$$

$$\frac{d^2y}{dx^2} = \frac{d}{dx}\left(\frac{dy}{dx}\right)\frac{d}{dx}(6x^2) = 12x.$$

Try this one:

$$y = x + \frac{1}{x}$$

$$\frac{d^2y}{dx^2} = [-\frac{1}{x^2} \mid \frac{1}{x} \mid +\frac{2}{x^3} \mid \text{none of these}]$$

If right, go to **245**.
If wrong, go to **244**.

Answer: (238) $\frac{-1}{x(\ln x)^2}$

244

Here is the solution to **243**.

$$y = x + \frac{1}{x},$$

$$\frac{dy}{dx} = 1 - \frac{1}{x^2},$$

$$\frac{d^2y}{dx^2} = 0 - 1\left(\frac{-2}{x^3}\right) = \frac{2}{x^3}.$$

Go to **245**.

245

An example of a second derivative with which you may already be familiar is *acceleration*.

Velocity is the rate of change of position with respect to time.

$$v = \frac{dS}{dt}.$$

Acceleration a is the rate of change of velocity with respect to time. Hence

$$a = \frac{dv}{dt}.$$

It follows then that

$$a = \frac{d}{dt}\left(\frac{dS}{dt}\right) = \frac{d^2S}{dt^2}.$$

Go to **246**.

246

The position of a particle is given by

$$S = A \sin \omega t.$$

A and ω (omega) are constants. Find the acceleration.

Answer: [0 | $A\omega \cos \omega t$ | $(A\omega \cos \omega t)^2$ | $-A\omega^2 \sin \omega t$].

If right, go to **248**.
If wrong, go to **247**.

247

$$\text{Acceleration} = \frac{d^2S}{dt^2} = \frac{d^2}{dt^2}(A \sin \omega t).$$

$$\frac{dS}{dt} = \frac{d}{dt}(A \sin \omega t) = A\omega \cos \omega t \qquad \text{(see frame } \mathbf{219}\text{)},$$

$$\frac{d^2S}{dt^2} = \frac{d}{dt}\left(\frac{dS}{dt}\right) = \frac{d}{dt}(A\omega \cos \omega t) = -A\omega^2 \sin \omega t.$$

Go to **248**.

248

There is really nothing essentially new about a second derivative. In fact, we can define derivatives of any order n, where n is a positive integer. Thus, $\frac{d^nf}{dx^n}$ is the nth derivative of f with respect to x. Try this problem:

If $f = x^4$, find $\frac{d^4y}{dx^4}$.

Answers: (242) $12x$ **(243)** $\frac{2}{x^3}$

$$\frac{d^4f}{dx^4} = [x^{16} \mid 4x^4 \mid 0 \mid 64 \mid 4 \times 3 \times 2 \times 1]$$

Go to **249**.

249

$$\frac{d^4}{dx^4}(x^4) = \frac{d}{dx}\left(\frac{d}{dx}\left\{\frac{d}{dx}\left[\frac{d}{dx}(x^4)\right]\right\}\right)$$

$$= \frac{d^3}{dx^3}(4x^3) = \frac{d^2}{dx^2}(4 \times 3x^2) = \frac{d}{dx}(4 \times 3 \times 2x)$$

$$= 4 \times 3 \times 2 \times 1.$$

We can easily generalize this result:

$$\frac{d^n}{dx^n}(x^n) = n \times (n-1) \times (n-2) \times \cdots \times 1$$

$$= n!$$

[$n!$ is called n factorial and is $n \times (n-1) \times (n-2) \times \cdots \times 1$.]

For more practice on higher-order derivatives, see problems 59 through 63 on page 248.

Go on to **250**.

Maxima and Minima

250

Now that we know how to differentiate simple functions, let's put our knowledge to use. Suppose we want to find the value of x and y at which

$$y = f(x)$$

has a minimum or a maximum value in some given region. By the end of this section we will know how to solve this problem.

Go to **251**.

251

Here is the graph of a function. At which of the points indicated does y have a minimum value in the domain plotted?

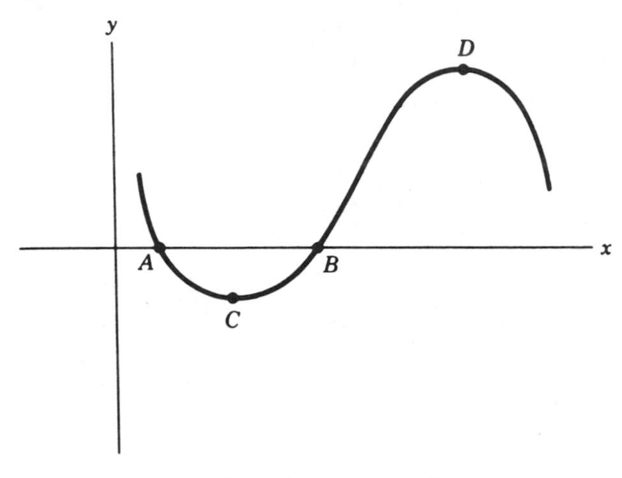

[A | B | C | D | A and B | C and D]

If correct, go to **253**.
If wrong, go to **252**.

252

The minimum value of y is at point C only, since y has its smallest value at point C, at least for the domain of x plotted.

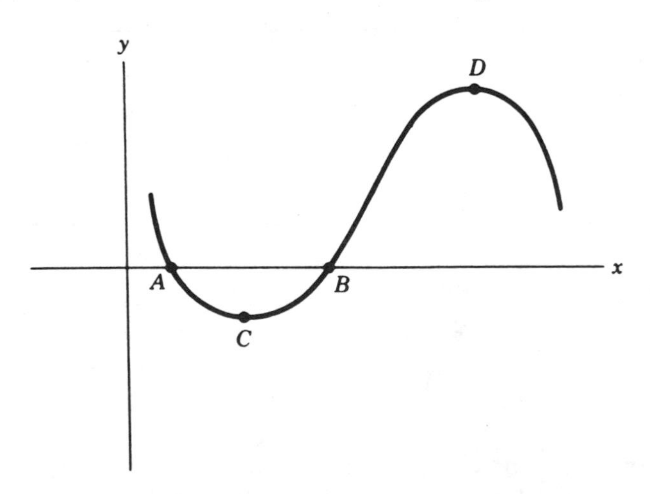

Answers: **(246)** $-A\omega^2 \sin \omega t$
(248) $4 \times 3 \times 2 \times 1$

At A and B, y has the value 0, but this has nothing to do with whether or not it has a minimum value there.

Point D is a *maximum* value of y.

Go to **253**.

253

We have shown that point C corresponds to a minimum value of y, at least insofar as nearby values are concerned, and that D corresponds similarly to a maximum value.

There is an interesting relation between the points of maximum or minimum values of y and the value of the derivative at those points. To help see this, sketch a plot of the derivative of the function shown, using the space provided.

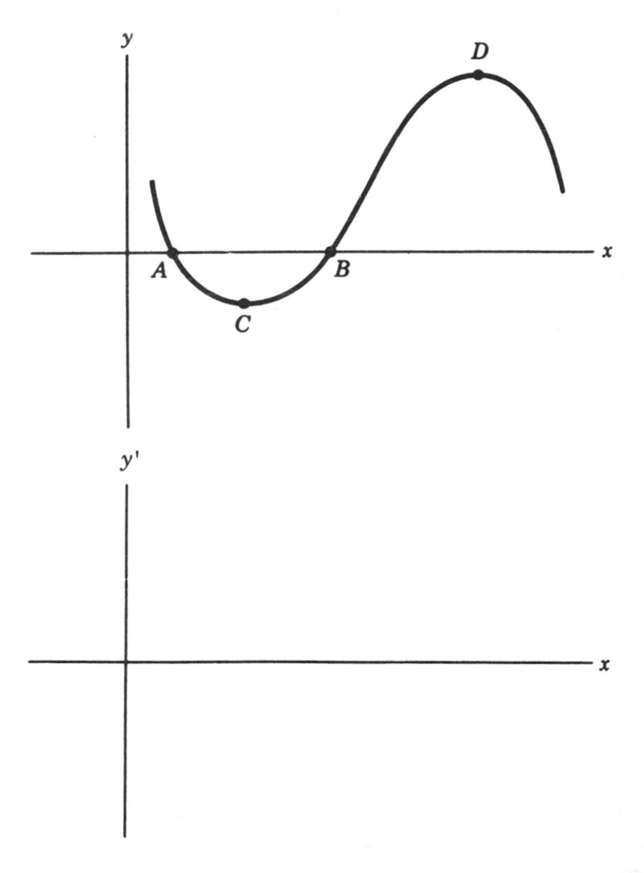

To check your sketch,

Go to **254**.

254

If you did not obtain a sketch substantially like this, review frames **160** to **169** before continuing.

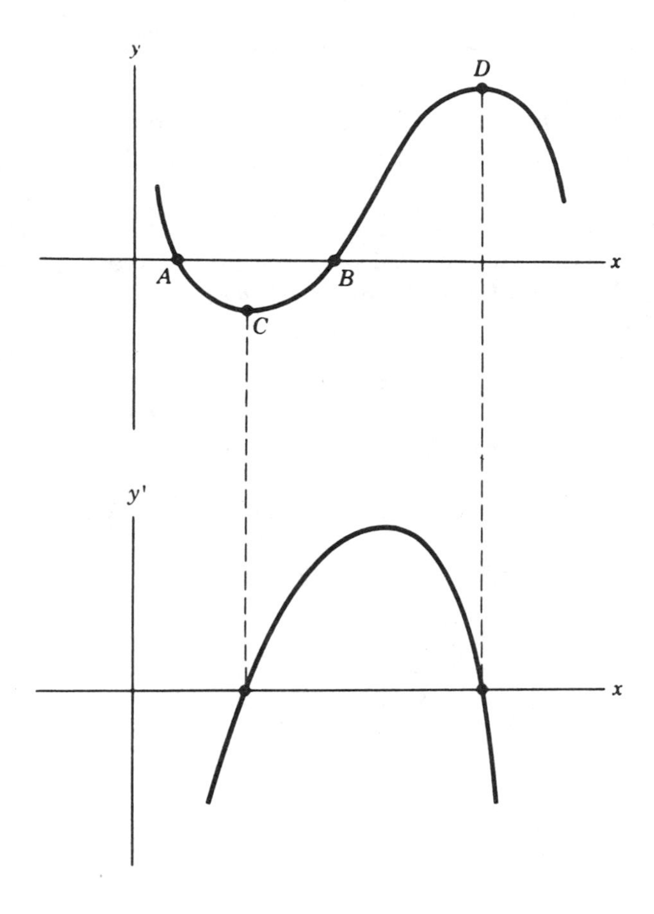

This simple example should be enough to convince you that if $f(x)$ has a maximum or a minimum for some value of x within a given interval, then its derivative f' is zero for that x.

One way to tell whether it is a maximum or a minimum is to plot a few neighboring points. However, there is an even simpler method, as we shall soon see.

Go to **255**.

Answer: (251) C

255

Test yourself with this problem:

Find the value of x for which the following has a minimum value.

$$f(x) = x^2 + 6x.$$

[-6 | -3 | 0 | $+3$ | none of these]

If right, go to **258**.
If wrong, go to **256**.

256

The problem is solved as follows:

The maximum or minimum occurs where x satisfies $f' = 0$.

$$f(x) = x^2 + 6x, \qquad f' = 2x + 6.$$

Thus the equation for the value of x at the maximum or minimum is

$$2x + 6 = 0 \qquad \text{or} \qquad x = -3.$$

Here is another problem:

For which values of x does the following $f(x)$ have a maximum or minimum value?

$$f(x) = 8x + \frac{2}{x}.$$

[$\frac{1}{4}$ | $-\frac{1}{4}$ | -4 | 2 and -4 | $\frac{1}{2}$ and $-\frac{1}{2}$]

If you were right, go to **258**.
If you did not get the correct answer, go to **257**.

257

The problem in frame **256** can be solved as follows:

At the position of maximum or minimum, $f' = 0$. Since

$$f(x) = 8x + \frac{2}{x}, \qquad f' = 8 - \frac{2}{x^2}.$$

The desired points are solutions of

$$8 - \frac{2}{x^2} = 0 \qquad \text{or} \qquad x^2 = \frac{2}{8} = \frac{1}{4}.$$

Thus at $x = +\frac{1}{2}$ and $x = -\frac{1}{2}$, $f(x)$ has a maximum or a minimum value. A plot of $f(x)$ is shown in the figure, and, as you can see, $x = -\frac{1}{2}$ yields a maximum, and $x = +\frac{1}{2}$ yields a minimum.

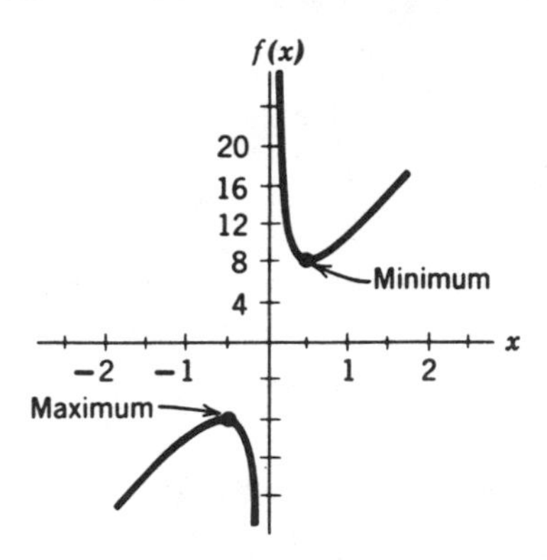

Incidentally, as you can see from the drawing, the minimum falls above the maximum. This should not be paradoxical, since we are talking about local minima or maxima—that is, the minimum or maximum value of a function in some small region.

Go to **258**.

258

We mentioned earlier that there is a simple method for finding whether $f(x)$ has a maximum or a minimum value when $f' = 0$. Let's find the method by drawing a few graphs.

Answers: (255) −3 **(256)** ½ and −½

Below are graphs of two functions. On the left, $f(x)$ has a maximum value in the region shown. On the right, $g(x)$ has a minimum value. In the spaces provided, draw rough sketches of the derivatives of $f(x)$ and $g(x)$.

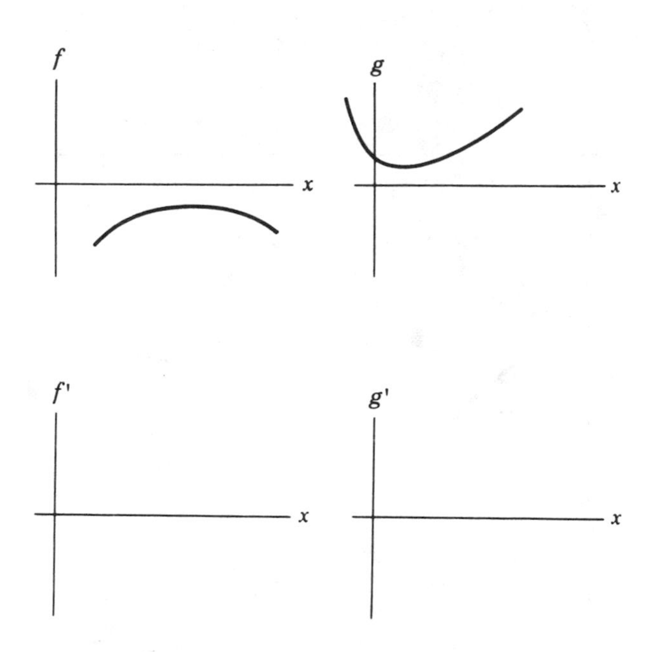

Now, let's repeat the process again. Make a rough sketch of the *second* derivative of each function (i.e., sketch the derivatives of the new functions you have just drawn).

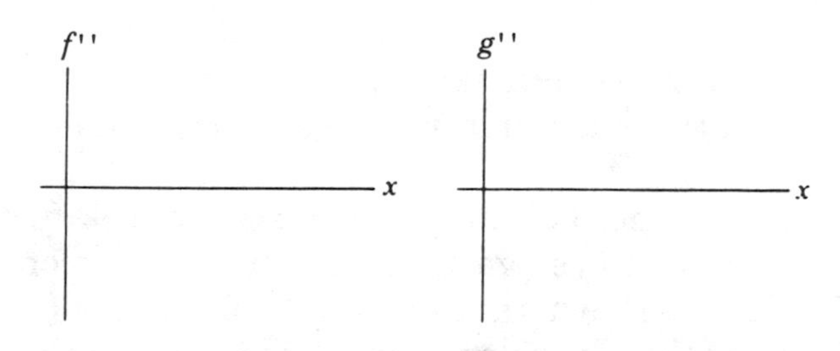

Perhaps from these sketches you can guess how to tell whether the function has a maximum or a minimum value when its derivative is 0. Whether you can or not,

Go to **259**.

259

The sketches should look approximately like this.

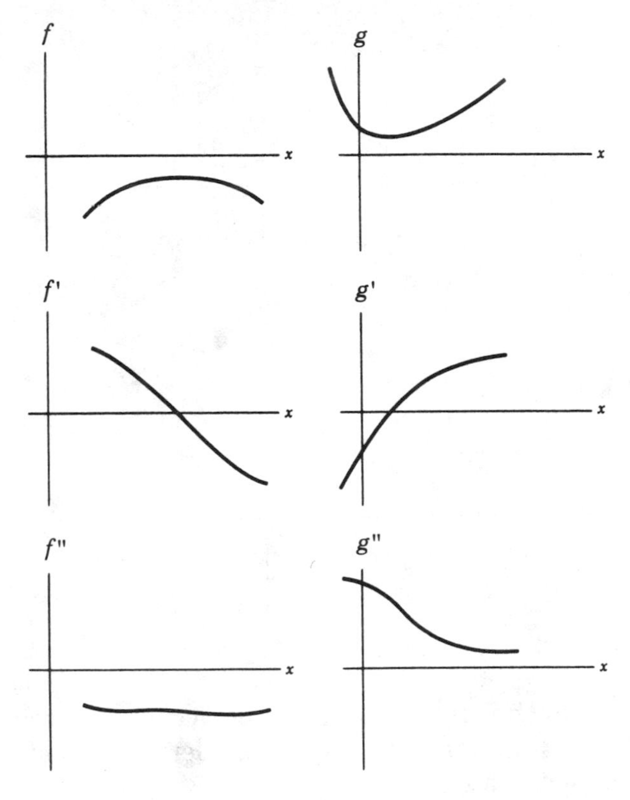

By studying these sketches, it should become apparent that wherever $f' = 0$,

$$f(x) \text{ has a maximum value if } f'' < 0,$$
$$\text{and } f(x) \text{ has a minimum value if } f'' > 0.$$

(If $f'' = 0$, this test is not helpful and we have to look further.)

If you are not convinced yet, go back and sketch the second derivatives of any of the functions shown in frames **164**, **166**, or **168** [(c) or (d)]. This should convince you that the rule is reasonable. Whenever you are ready,

Go to **260**.

260

Here is one last problem to try before we go on to another subject. Consider $f(x) = e^{-x^2}$. Find the value of x for which $f(x)$ has a maximum or minimum value, and determine which it is.

Answer: ______________

To check your answer, go to **261**.

261

Let's solve the problem: $f(x) = e^{-x^2}$. Using the chain rule, we find

$$f' = -2xe^{-x^2}.$$

Maximum or minimum occurs at x given by

$$-2xe^{-x^2} = 0 \qquad \text{or} \qquad x = 0.$$

Now we use the product rule (frame **189**) to get

$$f'' = -2e^{-x^2} + 4x^2e^{-x^2} = (-2 + 4x^2)e^{-x^2}.$$

At $x = 0, f'' = (-2 + 4 \times 0) \times 1 = -2$. Since f'' is negative where $f' = 0$, at $x = 0$, $f(x)$ has a *maximum* value there.

A word of caution—in evaluating a derivative, say f' at some value of $x, x = a$, you must always *first differentiate* $f(x)$ and then substitute $x = a$. If you reverse the procedure and first evaluate $f(a)$ and then try to differentiate it, the result will simply be 0 since $f(a)$ is a constant. Similar care must be taken with higher-order derivatives.

Go on to **262**.

Differentials

262

So far we have denoted the derivative by the symbol y' or $\frac{dy}{dx}$. Although either symbol stands for $\lim_{\Delta x \to 0} \frac{\Delta y}{\Delta x}$, the method of writing $\frac{dy}{dx}$ suggests that the derivative might be regarded as the ratio of two quantities, dy and dx. This turns out to be the case. The new quantities which we now introduce are called *differentials,* and they are defined in the next frame.

Go on to **266**.

263

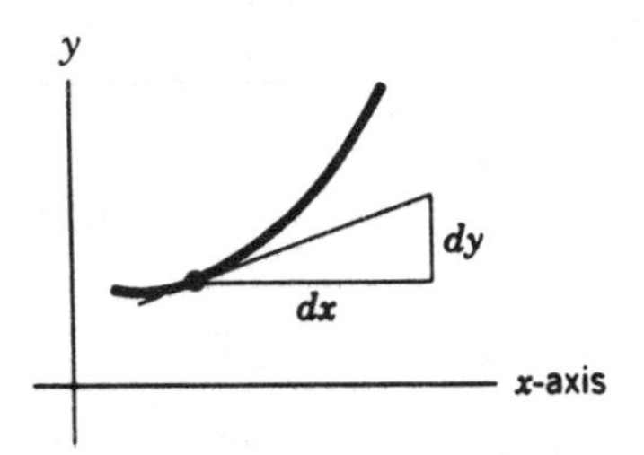

Suppose that x is an independent variable, and that $y = f(x)$. Then the *differential* dx of x is defined as equal to any increment, $x_2 - x_1$, where x_1 is the point of interest. The differential dx can be positive or negative, large or small, as we please. We see that dx, like x, can be regarded as an independent variable. The differential dy is defined by the following rule:

$$dy = y' \, dx,$$

where y' is the derivative of y with respect to x.

Go to **264**.

264

Although the meaning of the derivative y' is $\lim_{\Delta x \to 0} \frac{\Delta y}{\Delta x}$, we can see from the preceding frame that it can now be interpreted as the ratio of the differentials dy and dx, where dx is any increment of x and dy is *defined* by the rule $dy = y' \, dx$.

Go to **265**.

265

It is important not to confuse dy with Δy. As was pointed out in frame **136**, Δy stands for $y_2 - y_1 = f(x_2) - f(x_1)$, where x_2 and x_1 are two given values of x. Both dx and Δx $(=x_2 - x_1)$ are arbitrary intervals. dx is called a *differential* of x, and Δx is called an increment of x, but their meanings are similar here.

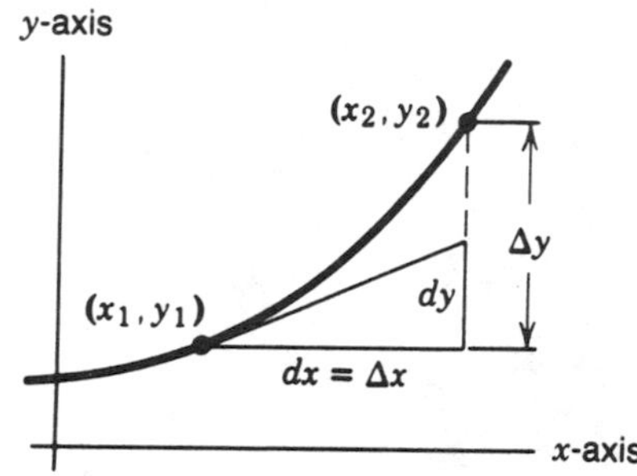

The diagram should show that dy and Δy are different quantities. Here we have set $dx = \Delta x$. The *differential* dy is then $y'\,dx$, while the *increment* Δy is given by $y_2 - y_1$. It is clear in this case that dy is not the same as Δy.

Go to **266**.

266

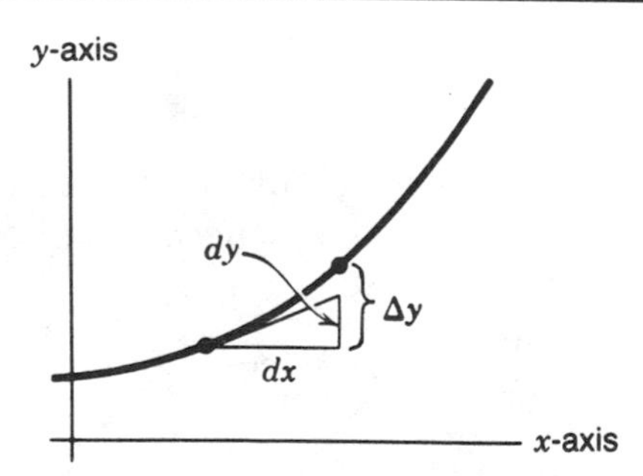

Although dy and Δy are different, you can see from the figure that for sufficiently small dx (with $dx = \Delta x$) dy is very close to Δy. We can write this symbolically as

$$\lim_{dx=\Delta x \to 0} \frac{dy}{\Delta y} = 1.$$

Hence, if we intend to take the limit where $dx \to 0$, dy may be substituted for Δy. Furthermore, even if we don't take the limit, dy is almost the same as Δy, provided dx is sufficiently small. We, therefore, often use dy and Δy interchangeably when it is understood that the limit will be taken or that the result may be an approximation.

Go to frame **267**.

267

We can rewrite in differential form the various expressions for derivatives given earlier. Thus, if $y = x^n$,

$$dy = d(x^n) = \frac{d}{dx}(x^n)\,dx = nx^{n-1}\,dx.$$

Find the following:

$$d(\sin x) = [-\sin x\,dx \mid -\sin x \mid -\cos x\,dx \mid \cos x\,dx]$$

$$d\left(\frac{1}{x}\right) = [\frac{dx}{x^2} \mid -\frac{dx}{x^2} \mid -\frac{dx}{x}]$$

$$d(e^x) = [xe^x\,dx \mid dx \mid e^x\,dx \mid \frac{dx}{e^x}]$$

If you missed any of these go to **268**.
Otherwise, go to **269**.

268

Here are the solutions to the problems in frame **267**. The number of the frame in which each derivative is discussed is shown in parentheses.

$$d(\sin x) = \left(\frac{d \sin x}{dx}\right) dx = \cos x\,dx \qquad \text{(frame 211)},$$

$$d\left(\frac{1}{x}\right) = \left[\frac{d}{dx}\left(\frac{1}{x}\right)\right] dx = -\frac{dx}{x^2} \qquad \text{(frame 180)},$$

$$d(e^x) = \left(\frac{d}{dx}(e^x)\right) dx = e^x\,dx \qquad \text{(frame 235)}.$$

Go to **269**.

269

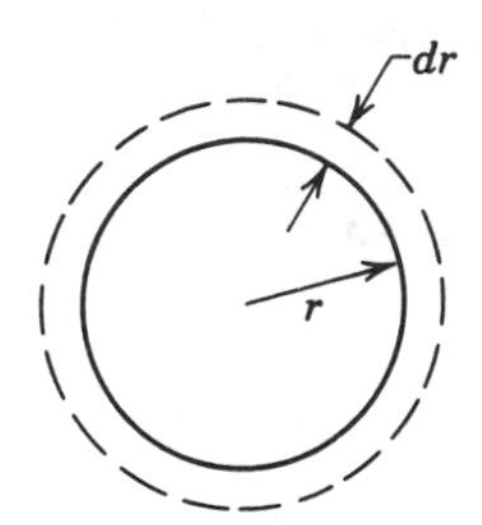

Here is an example of the use of a differential. The diagram shows the surface of a disc to which a thin rim has been added. Suppose we want an approximate value for the change area dA which occurs when the radius is increased from r to $r + dr$.

$$dA = \left(\frac{dA}{dr}\right) dr = \frac{d}{dr}(\pi r^2)\, dr = 2\pi r\, dr.$$

Go to **270**.

270

The previous example can also be solved exactly by taking the difference of the two areas:

$$\Delta A = \pi(r + \Delta r)^2 - \pi r^2 = 2\pi r\, \Delta r + \pi\, \Delta r^2.$$

When Δr is small compared with r, we can neglect the last term and we see that

$$\Delta A = 2\pi r\, \Delta r.$$

If we let $\Delta r = dr$ and assume that they are both small then, as we know from frame **269**,

$$dA \approx \Delta A = 2\pi r\, dr.$$

(continued)

Here is a more intuitive argument for the results. Since the rim is thin, its area dA is the approximate length, $2\pi r$, multiplied by its width, dr. Hence,

$$dA = 2\pi r\, dr.$$

Go to **271**.

271

Differentials are handy for remembering some important rules for differentiation. For instance, the chain rule

$$\frac{dw}{dx} = \frac{dw}{du}\frac{du}{dx}$$

is almost an identity if we treat dw, du and dx as differentials. Actually, it is not obvious that we can do so, since w and u both depend on a third quantity, x. Justification for using differentials to obtain the chain rule is given in Appendix A9.

Go to **272**.

272

Here is another relation which is easy to remember with differentials, though the actual proof demands further explanation:

$$\boxed{\frac{dx}{dy} = \frac{1}{dy/dx}.}$$

This handy rule lets us reverse the role of dependent and independent variables, though it holds true only under certain conditions. If you want a further explanation, see Appendix A10.

Otherwise, go to **273**.

Answers: (267) $\cos x\, dx$, $-\dfrac{dx}{x^2}$, $e^x\, dx$

A Little Review and a Few Problems

273

Let's end the chapter by reviewing some of the ideas introduced early in the chapter and by putting differential calculus to work in a few problems involving velocity.

Go to **274**.

274

We hope you recall that the rate of change of position of a moving point with respect to time is called the velocity.

In other words, if position and time are related by a function S, in order to find the velocity, we ________________ $S(t)$ with respect to ________________.

Go to **275**.

275

You should have written

In other words, if the position and time are related by a function S, in order to find the velocity, we *differentiate* $S(t)$ with respect to *time* (or t).

Go to **276**.

276

Can you answer this problem?

The position of a particle along a straight line is given by the following expression:

$$S = A \sin \omega t.$$

A and ω (omega) are constants.

Find the velocity of the particle.

$$v = \underline{\hspace{4cm}}.$$

For the answer, go to **277**.

277

Your answer should have been

$$v = A\omega \cos \omega t.$$

If you got the right answer, skip on to **280**. Otherwise, continue here.

The problem is to find the velocity, which is the rate of change of position with respect to time.

In this problem, the position is $S = A \sin \omega t$.

$$v = \frac{dS}{dt} = \frac{d}{dt}(A \sin \omega t) = A\omega \cos \omega t.$$

(If you are not sure of the procedure here, see frame **219**.)

Can you do this problem?

$$S = A \sin \omega t + B \cos 2\omega t.$$

Find v.

$$v = \_\_\_\_\_\_\_\_\_\_\_\_\_\_\_\_.$$

See frame **278** for the answer.

278

$$v = \frac{d}{dt}(A \sin \omega t + B \cos 2\omega t)$$

$$= A\omega \cos \omega t - 2B\omega \sin 2\omega t.$$

If you wrote this, go to **280**. If not, review frame **220** and then continue here.

Try this problem: The position of a point is given by

$$S = A \sin \omega t \cos \omega t.$$

Find its velocity.

$$v = \_\_\_\_\_\_\_\_\_\_\_\_\_\_\_\_.$$

Go to **279** for the answer.

279

Here is how to solve problem **278**.

$$v = \frac{dS}{dt} = \frac{d}{dt}(A \sin \omega t \cos \omega t)$$

$$= A \sin \omega t \frac{d}{dt}(\cos \omega t) + A \left[\frac{d}{dt}(\sin \omega t)\right]\cos \omega t$$

$$= -A\omega \sin^2 \omega t + A\omega \cos^2 \omega t$$

$$= A\omega(\cos^2 \omega t - \sin^2 \omega t.$$

As an alternative approach you might note that

$$\sin \omega t \cos \omega t = \tfrac{1}{2}(\sin 2\omega t).$$

(See frame **71**.) Then, $v = \frac{d}{dt}\left(\frac{A}{2} \sin 2\omega t\right)$. If you feel energetic, show that this procedure yields the same result as above.

Go to **280**.

280

Suppose the height of a ball above the ground is given by $y = a + bt + ct^2$ where a, b, c, are constants. (Here we are using y rather than S to denote position. It makes no difference what we call our variable. This type of equation actually describes the height of a freely falling body.)
Find the velocity in the y direction.

$v =$ ____________________.

See **281** for the correct answer.

281

Here is how to do the problem in frame **280**.

$$v = \frac{dy}{dt} = \frac{d}{dt}(a + bt + ct^2) = b + 2ct.$$

If you wrote the correct answer, go to **283**. Otherwise, do the problem below. Let

$$S = \frac{e}{t^2} + bt \qquad (e \text{ and } b \text{ are constants}).$$

(continued)

Find the velocity.

$$v = \underline{\hspace{6cm}}$$

The answer is in frame **282**.

282

$$v = \frac{dS}{dt} = \frac{d}{dt}\left(\frac{e}{t^2} + bt\right) = -\frac{2e}{t^3} + b.$$

If this problem gave you any difficulty you should review the beginning of this section before going on.

Otherwise, go to **283**.

283

Here is a more difficult problem which you may enjoy. (If you don't feel in the mood, skip on to frame **285**.)

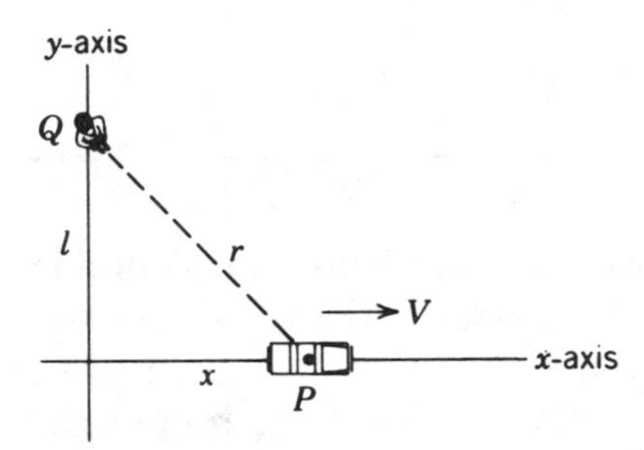

A car P moves along a road in the x direction with a constant velocity V. The problem is to find how fast it is moving away from a man standing at point Q, distance l from the road, as shown. In other words, if r is the distance between Q and P, find $\frac{dr}{dt}$.

$\left(\textit{Hint}\right.$: The chain rule is very useful here in the form $\left.\frac{dr}{dt} = \frac{dr}{dx}\frac{dx}{dt}.\right)$

$$\frac{dr}{dt} = \underline{\hspace{6cm}}.$$

Go to **284** after working this out.

284

From the diagram in **283** you can see that

$$r^2 = x^2 + l^2, \qquad r = (x^2 + l^2)^{1/2}.$$

We must find $\frac{dr}{dt}$, and we can do so in the following way:

$$\frac{dr}{dt} = \frac{dr}{dx}\frac{dx}{dt} = \frac{d}{dx}(x^2 + l^2)^{1/2}\frac{dx}{dt}$$

$$= \frac{1}{2}\frac{2x}{(x^2 + l^2)^{1/2}}\frac{dx}{dt}$$

$$= V\frac{x}{(x^2 + l^2)^{1/2}}.$$

In the last step we have used $V = \frac{dx}{dt}$.

Go to **285**.

285

The problem is to maximize the gross income from selling a new electronic entertainment device, a Home Whoosie. If S Whoosies are sold at price x dollars, the income is Sx. However, as the price is raised the number of buyers decreases. It is estimated that the number of buyers at price x can be described by the following expression: $S(x) = S_0[1-(x/x_0)^2]$, where S_0 and x_0 are constants. Note that if $x << x_0$, the number of sales is practically independent of price, but that as the price is increased and x approaches x_0 the sales rapidly fall, vanishing at x_0. (The expresssion is meaningless for $x > x_0$.)

What should be the price x for the maximum income, and what is he maximum gross income I?

$x =$ ________________.

$I =$ ________________.

Go to **286**.

286

The gross income is

$$I = Sx = S_0[1 - (x/x_0)^2x]$$
$$= S_0(x - x^3/x_0^2).$$

The maximum value of I occurs when $\dfrac{dI}{dx} = 0.$

$$\frac{dI}{dx} = S_0(1 - 3x^2/x_0^3) = 0$$

$$x^2 = x_0^2/3, \qquad \text{hence } x = \sqrt{1/3}\,x_0 = 0.577x_0.$$

At the maximum, the gross income is

$$I = Sx = S_0[1 - (x/x_0)^2]x = S_0\,(1 - 1/3)\sqrt{1/3}\,x_0$$
$$= 0.385S_0I_0.$$

The conclusion is that the price x should be set at about 57.7 percent of the maximum value, x_0 dollars, and that the gross income is about $0.385S_0$ dollars.

These particular conclusions hold only for the price-sales curve, $S(x) = S_0[1-(x/x_0)^2]$. However, the method used here can identify the price that gives the maximum profit for any price-sales curve you wish to choose.

Go to **287**.

287

In deciding whether or not to keep an old automobile, an important consideration is the estimated cost per year of owning the car. The two major components of the cost are repairs and depreciation. We shall assume that the annual repairs cost r, in dollars per year, is given by

$$r = A + Bt,$$

where A and B are constants. The repairs are lowest when the car is new and are assumed to increase linearly with time. The rate of depreciation—the loss in value of the car in dollars per year—is taken to be

$$d = D_0e^{-ct},$$

where D_0 and c are constants. The depreciation rate is highest when the car is new and most valuable; it decreases exponentially in time, growing smaller as the car becomes less valuable.

The annual cost due to repairs and depreciation is $S = r + d$. Find an expression for the time t at which the cost is a minimum.

Time = ______________.

Go to **288**.

288

The cost is

$$S = r + d = A + Bt + De^{-ct}.$$

An extremum occurs when $dS/dt = B - cDe^{-ct} = 0$. This can be solved for t:

$$cDe^{-ct} = B, \qquad e^{-ct} = \frac{B}{cD},$$

$$-ct = \ln \frac{B}{cD},$$

$$t = \frac{1}{c} \ln \frac{cD}{B}.$$

To see whether this is a minimum or a maximum, we must examine d^2S/dt^2. (Recall form frame **259** that the second derivative is positive at a minimum.)

$$\frac{d^2S}{dt^2} = c^2De^{-ct}.$$

This is always positive, so the extremum is a minimum. Note, however, that if $cD/B < 1$, then $\ln(cD/B) < 0$, and t is negative. What does this mean?

Consider dS/dt at $t = 0$.

$$\left.\frac{dS}{dt}\right|_{t=0} = B - cDe^0 = B - cD.$$

(continued)

If $B < cD$, the slope is negative. S initially decreases and has a minimum at some later time just before it starts to increase. If $B > cD$, however, S increases at $t = 0$ and keeps on increasing. This is the case for which the minimum in S occurs at a negative time. This solution has no meaning; you can't sell a car before you have bought it!

Go to **289**.

Conclusion to Chapter 2

289

The Appendixes contain additional material which may be helpful. For instance, sometimes one has an equation which relates two variables, y and x, but which cannot be written simply in the form $y = f(x)$. There is a straightforward method for evaluating y': it is called *implicit differentiation* and it is described in Appendix B1. Appendix B2 shows how to differentiate the inverse trigonometric functions. In this chapter we have only discussed differentiation of functions of a single variable. It is not difficult to extend the ideas to functions of several variables. The technique for doing this is known as *partial differentiation*. If you are interested in the subject, see Appendix B3.

Many problems involving rates of change lead to equations which express relations between functions and their derivatives. Such relations are known as *differential equations*. Appendix B4 discusses two common types of differential equations.

All the important results of this chapter are summarized in Chapter 4. You may wish to read that material now as a quick review. In addition, a list of important derivatives is presented in Table 1 at the back of the book.

Don't forget the review problems, page 246, if you want more practice.

Ready for more? Take a deep breath and go on to Chapter 3.

CHAPTER THREE
Integral Calculus

We are now ready to tackle integral calculus. In this chapter you will learn:

- About antidifferentiation and indefinite integrals
- The meaning of integration
- How to find the area under curves
- How to evaluate definite integrals
- How to integrate numerically
- Some applications of integral calculus
- How to use multiple integrals

The Area under a Curve

290

In this chapter we are going to learn about the second major branch of calculus—integral calculus. The first branch—differential calculus—stems from the problem of finding the *slope* of the graph of a function.

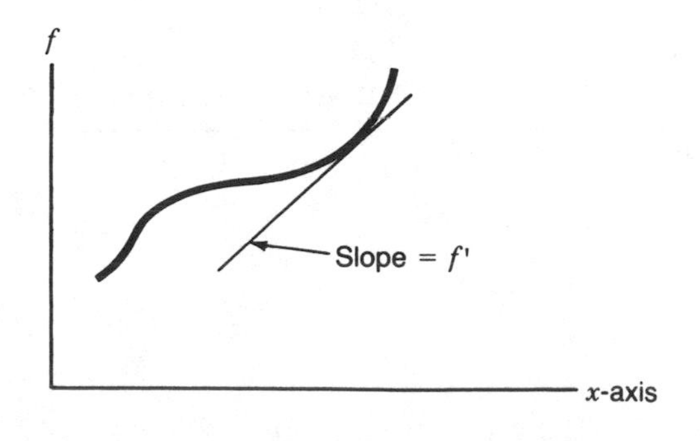

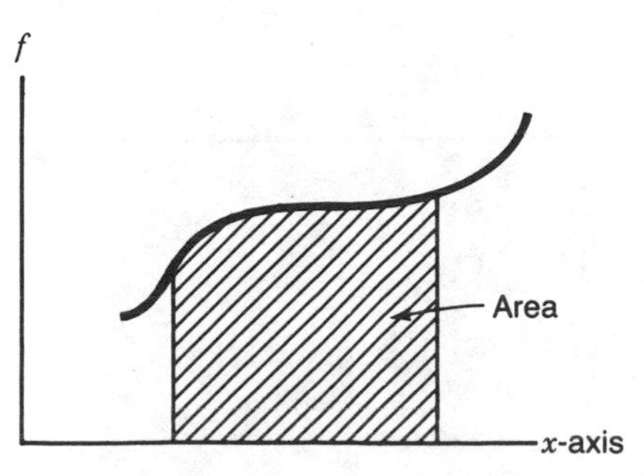

(continued)

Integral calculus stems from another problem related to the graph of $f(x)$: how to calculate the area between $f(x)$ and the x-axis bounded by some initial point a and some arbitrary final point, x, as shown in the drawing. The area can be calculated by a process called *integration*. Just as differentiation is useful for many applications besides finding slopes of curves—for instance, calculating rates of growth or finding maxima and minima—so integral calculus has many applications besides finding areas under curves. However, the area problem motivated the creation of integral calculus, and we shall use it to motivate the explanation of integration.

Go to **291**.

291

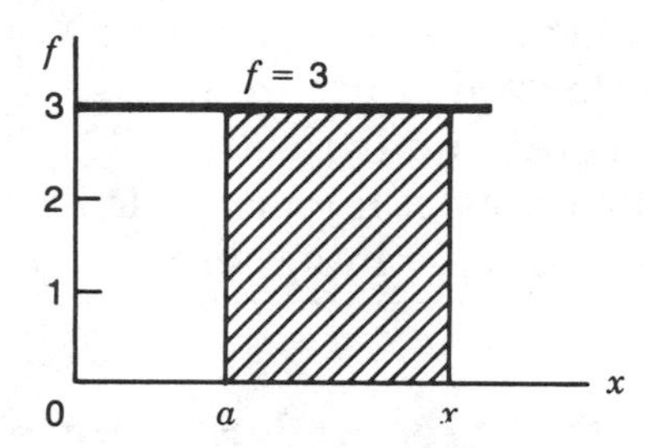

To illustrate what is meant by the area under a curve, here is a graph of the simplest of all curves—a straight line given by $f(x)$ = constant. What is the area $A(x)$ between the line $f(x) = 3$, and the x-axis between the interval a and some arbitrary point $x > a$?

$$A(x) = [3ax \mid 3(a + x) \mid 3(a - x) \mid 3(x - a)]$$

To check your answer, go to **292**.

292

The area in the rectangle is the product of the base, $x - a$, and the height, 3. Thus the area is $3(x - a)$.

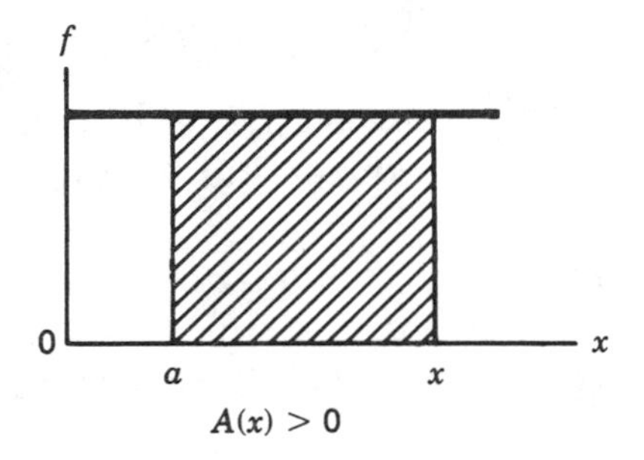

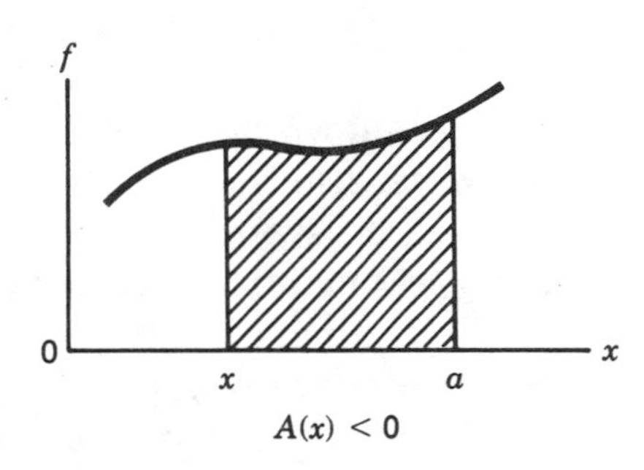

There is an important convention in the sign of the area, for areas can be positive or negative. In the drawing at the left, $x - a$ is positive since $x > a$, and $f(x)$ is also positive. Because the base and height are both positive, the area is positive. However, the area under the graph at the right is negative, since the base is $x - a < 0$. Thus, areas can be positive or negative.

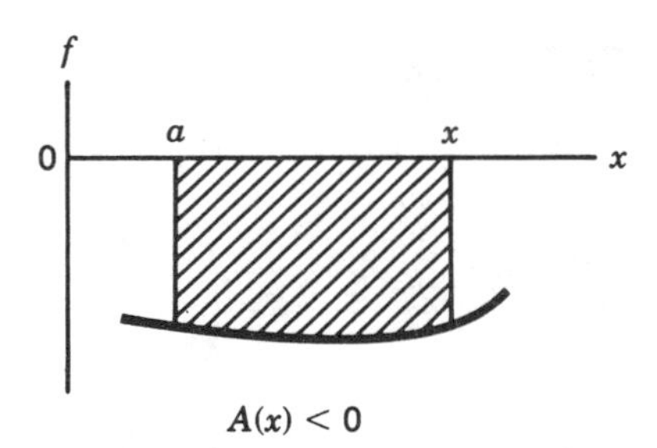

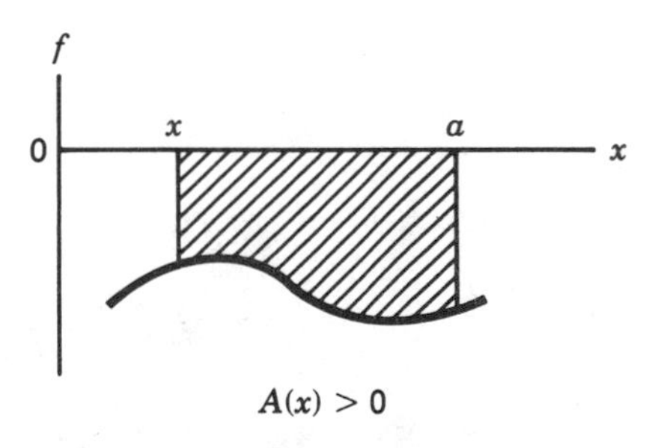

If $f(x)$ is negative while the base interval is positive, the area (base × height) is negative, whereas if both $f(x)$ and the base are negative, the area is the product of two negative numbers and is positive.

Go to **293.**

293

The area $A(x)$ and the function $f(x)$ are closely related; the *derivative* of the area is simply $f(x)$.

$$\boxed{A'(x) = f(x).}$$

We will explain why this relation is true in the next section, but for the present, we shall simply use it. To find $A(x)$, we must find a function which, when differentiated, is $f(x)$.

Finding a function which, when differentiated, yields another function, is often called *antidifferentiation*. The term is descriptive, for the process essentially involves "differentiation backwards." A more formal term for the process is *integration*.

Go to **294**.

294

To illustrate that $A'(x) = f(x)$, let's look at some simple areas that one can calculate directly. We have already discussed the area under the curve $f(x) = C$, where C is a constant,

$$A(x) = C(x - a).$$

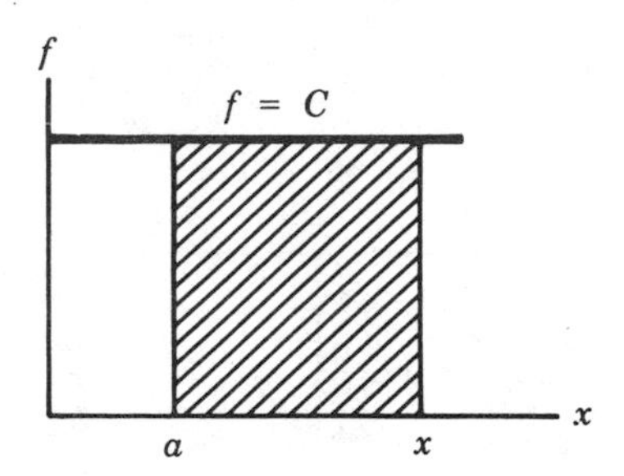

Differentiating,

$$A'(x) = C = f(x).$$

Find the area $A(x)$ under $f(x) = Dx$ between a and x, and prove to yourself that $A'(x) = f(x)$.

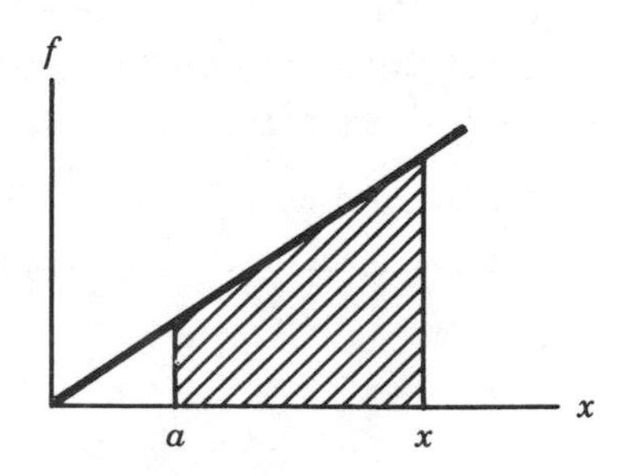

If you want to check your result, go to **295**.
Otherwise, skip to **296**.

295

One way to calculate the area is to think of it as the difference of the area of two right triangles. Using area = ½ base × height, we have

$$A(x) = \frac{1}{2}xf(x) - \frac{1}{2}af(a) = \frac{1}{2}Dx^2 - \frac{1}{2}Da^2,$$

$$A'(x) = \frac{d}{dx}\left(\frac{1}{2}Dx^2 - \frac{1}{2}Da^2\right) = Dx = f(x).$$

Go to **296**.

296

To see why $A'(x) = f(x)$, consider how the area $A(x)$ changes as x increases by an amount Δx. $A(x + \Delta x) = A(x) + \Delta A$, where ΔA is the thin strip shown.

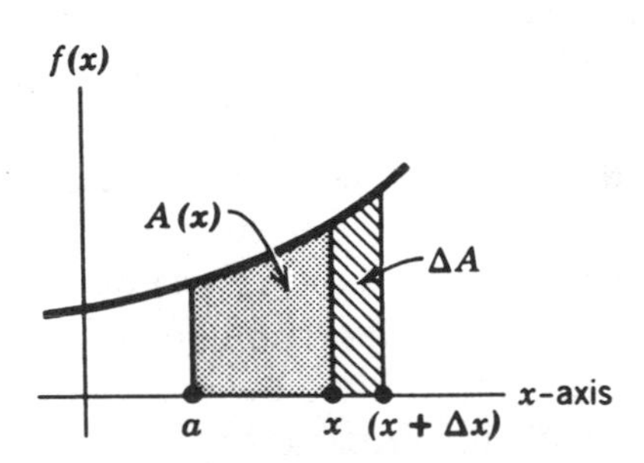

Can you find an approximate expression for ΔA?

$\Delta A \approx$ ____________________.

(≈ means "approximately equal to.")

This represents a major step in the development of integral calculus, so don't be disappointed if the result eludes you.

Go to **297**.

297

The answer we want is

$$\Delta A \approx f(x)\,\Delta x.$$

If you wrote this, go on to **298**. Otherwise, read below.

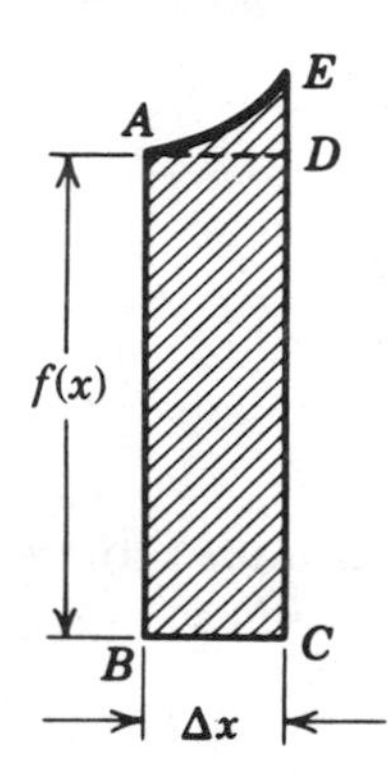

Let's take a close look at the area. As you can see, the area is a long thin strip. Unfortunately it is not a rectangle—but is nearly so. Most of the area is that of the rectangle $ABCD$ and this area is the product of the length $f(x)$ and its width Δx, that is, $f(x)\,\Delta x$. The desired area ΔA differs from the area of the rectangle by the area of the figure ADE, which is almost a triangle except that the side AE is not straight. When the value of Δx becomes smaller and smaller, the area of the figure ADE becomes smaller at an even more rapid rate because both its base AD *and* its height DE become smaller in contrast to the rectangle $ABCD$ for which the length $f(x)$ stays fixed and only the width, $BC = \Delta x$, decreases. [Perhaps this argument has a familiar ring to it. What we are implying is that the approximation approaches an equality in the *limit* where $\Delta x \to 0$. More precisely, $\lim_{\Delta x \to 0} \dfrac{\Delta A}{f(x)\,\Delta x} = 1$.]

For a sufficiently small value of Δx, then, we can say

$$\Delta A \approx f(x)\,\Delta x.$$

Note that with similar accuracy we could have said

$$\Delta A \approx f(x + \Delta x)\,\Delta x,$$

and with even greater accuracy

$$\Delta A \approx f\left(x + \frac{\Delta x}{2}\right)\Delta x.$$

However, the first of these is the simplest and is sufficiently accurate if Δx is small enough.

Go to **298**.

298

Using the approximate expression for ΔA, we can calculate $A'(x)$ and justify the identity $A'(x) = f(x)$.

$$A'(x) = \frac{dA}{dx} = \lim_{\Delta x \to 0} \frac{A(x + \Delta x) - A(x)}{\Delta x}$$
$$= \lim_{\Delta x \to 0} \frac{\Delta A}{\Delta x}.$$

If Δx is small we can use the result of frame **297**.

$$\Delta A \approx f(x)\,\Delta x.$$

As explained in the last frame, the approximation becomes more and more exact as Δx becomes smaller and smaller. Hence,

$$A'(x) = \lim_{\Delta x \to 0} \frac{\Delta A}{\Delta x} = \lim_{\Delta x \to 0} \frac{f(x)\,\Delta x}{\Delta x} = f(x).$$

Go to **299**.

299

To summarize this section, we have found that the area $A(x)$ under a curve defined by $y = f(x)$ satisfies the equation $A'(x) = f(x)$. Thus, if we can find a function whose derivative is $f(x)$, we can find the area.

As previously stated, the process of finding a function whose derivative is another function is called *integration* or *antidifferentiation*. In the following sections we will learn some methods for integration and apply them to finding areas and to other problems such as finding the distance traveled by a vehicle whose speed is changing or calculating the size of populations (or financial accounts) which are growing at a changing rate.

Go to **300**.

Integration

300

The goal of this section is to learn some techniques for *integration* or *antidifferentiation*. (We will use both terms now, but later drop the descriptive term *antidifferentiation* for the more commonly used term *integration*.)

In this section we will generally designate a function by $f(x)$, and its integral or antiderivative by $F(x)$. Thus

$$F'(x) = f(x).$$

This notation describes a basic property of the antiderivative, but it only defines the derivative $F'(x)$, not $F(x)$ itself.

The antiderivative is most frequently written in the following form:

$$F(x) = \int f(x)\,dx.$$

$\int f(x)\,dx$ is called the *indefinite integral* of $f(x)$, or simply the integral of $f(x)$. (Later we shall come to another type of integral, called a *definite integral*, which is a number, not a function of a variable.) The symbol $\int$ is called an integration symbol, but it is never used alone.

To summarize the notation, if $F'(x) = f(x)$, then $F(x)$ is the ______________ ______________ or the ______________ of $f(x)$.

Go to **301**.

301

Often one can find the antiderivative or indefinite integral simply by guesswork. For instance, if $f(x) = 1$, $F(x) = \int f(x)\,dx = x$. To prove this, note that

$$F'(x) = \frac{d}{dx}(x) = 1 = f(x).$$

However, x is not the only antiderivative of $f(x) = 1$; $x + c$, where c is a constant, is also an antiderivative because

$$\frac{d}{dx}(x + c) = 1 + 0 = f(x).$$

In fact, a constant can *always* be added to a function without changing its derivative. It is important not to omit this arbitrary constant; otherwise the answer is incomplete.

Go to **302**.

302

Since integration is the inverse of differentiation, for every differentiation formula in Chapter 2, there is a corresponding integration formula here. Thus from Chapter 2,

$$\frac{d \sin x}{dx} = \cos x,$$

so by the definition of the indefinite integral,

$$\int \cos x \, dx = \sin x + c.$$

Now you try one. What is

$$\int \sin x \, dx?$$

Answer: [$\cos x + c$ | $-\cos x + c$ | $\sin x \cos x + c$ | none of these]

Make sure you understand the correct answer (you can check the result by differentiation) and then

Go to **303**.

303

Now try to find these integrals (for simplicity, the constant c has been omitted from the answers).

(a) $\int x^n \, dx = \left[\frac{1}{n} x^n \;\middle|\; \frac{1}{n} x^{n+1} \;\middle|\; \frac{1}{n+1} x^{n+1} \;\middle|\; \frac{1}{n-1} x^n\right]$

(b) $\int e^x \, dx = \left[e^x \;\middle|\; xe^x \;\middle|\; \frac{1}{x} e^x \;\middle|\; \text{none of these}\right]$

If you did both of these correctly, you are doing fine and should skip to frame **305**.

If not, go to frame **304**.

304

If you missed these due to a careless mistake and if you now understand the problem, correct your mistake and go on to frame **305**. If not, review the definitions of the indefinite integral in the first section of this chapter and then continue here. If

$$F = \int f(x)\, dx,$$

then

$$\frac{dF}{dx} = f(x).$$

Therefore if we want to find F, we try to find an expression which when differentiated gives $f(x)$. Now the derivative of $\frac{x^{n+1}}{n+1}$ is given by

$$\frac{d}{dx}\left(\frac{x^{n+1}}{n+1}\right) = \frac{1}{n+1}\frac{dx^{n+1}}{dx} = \frac{1}{n+1}(n+1)\,x^n = x^n$$

by the formula for differentiating x^n in Chapter 2. Thus, including the integration constant c, we find $\int x^n\, dx = \frac{x^{n+1}}{n+1} + c$. (Note that this formula will not work for $n = -1$.)

Likewise, by Chapter 2,

$$\frac{d}{dx}(e^x) = e^x$$

so

$$\int e^x\, dx = e^x + c.$$

Go to **305**.

305

So far we have found integrals by looking for a function whose derivative is the integrand. Although this works well in many cases,

---

Answers: (300) Antiderivative, indefinite integral (in any order)

(302) $-\cos x + c$ **(303)** (a) $\frac{1}{n+1}x^{n+1}$, (b) e^x

especially after you have had some practice, it is helpful to have a list of some of the more important integrals. It is quite on the up and up to use such a list. If you make much use of calculus, you will eventually know by sight most of the integrals listed, or at least be sufficiently familiar with them to make a good guess at the integral. You can always check your guess by differentiation.

A table of important integrals is given in the next frame. You can check the truth of any one of the equations

$$\int f(x)\,dx = F(x)$$

by confirming that

$$\frac{dF(x)}{dx} = f(x).$$

We will shortly use this method to verify some of the equations.

Go to **306**.

306

List of Important Integrals. The arbitrary integration constant is omitted for simplicity; a and n are constants.

1. $\int a\,dx = ax$
2. $\int af(x)\,dx = a\int f(x)\,dx$
3. $\int (u + v)\,dx = \int u\,dx + \int v\,dx$
4. $\int x^n\,dx = \dfrac{x^{n+1}}{n + 1}, \qquad n \neq -1$
5. $\int \dfrac{dx}{x} = \ln x$
6. $\int e^x\,dx = e^x$
7. $\int e^{ax}\,dx = \dfrac{e^{ax}}{a}$
8. $\int b^{ax}\,dx = \dfrac{b^{ax}}{a \ln b}$
9. $\int \ln x\,dx = x \ln x - x$
10. $\int \sin x\,dx = -\cos x$

(continued)

11. $\int \cos x\, dx = \sin x$
12. $\int \tan x\, dx = -\ln(\cos x)$
13. $\int \cot x\, dx = \ln(\sin x)$
14. $\int \sec x\, dx = \ln(\sec x + \tan x)$
15. $\int \sin x \cos x\, dx = \frac{1}{2}\sin^2 x$
16. $\int \frac{dx}{a^2 + x^2} = \frac{1}{a}\tan^{-1}\frac{x}{a}$
17. $\int \frac{dx}{\sqrt{a^2 - x^2}} = \sin^{-1}\frac{x}{a}$
18. $\int \frac{dx}{\sqrt{x^2 \pm a^2}} = \ln(x + \sqrt{x^2 \pm a^2})$

For convenience this table is repeated as Table 2 near the back of the book (page 256).

Go to **307**.

307

Let's see if you can check some of the formulas in the table. Show that integral formulas 9 and 15 are correct.

If you have proved the formulas to your satisfaction, go to **309**.
If you want to see proofs of the formulas, go to **308**.

308

To prove that $F(x) = \int f(x)\,dx$, we must show that $\frac{dF(x)}{dx} = f(x)$.

9. $F(x) = x \ln x - x,\ f(x) = \ln x.$

$$\frac{dF}{dx} = \frac{d}{dx}(x \ln x - x) = x\left(\frac{1}{x}\right) + \ln x - 1 = \ln x = f.$$

15. $F(x) = \frac{1}{2}\sin^2 x,\ f(x) = \sin x \cos x = f.$

$$\frac{d}{dx}\left(\frac{1}{2}\sin^2 x\right) = \frac{1}{2}(2 \sin x)\frac{d}{dx}(\sin x) = \sin x \cos x.$$

Go to **309**.

Some Techniques of Integration

309

Often an unfamiliar function can be converted into a familiar function having a known integral by using a technique called *change of variable*. The method applies to integrating a "function of a function." (Differentiation of such a function was discussed in frame **198**. It is done using the chain rule.) For example, e^{-x^2} can be written e^{-u}, where $u = x^2$. With the following rule, the integral with respect to the variable x can be converted into another integral, often simpler, depending on the variable u.

$$\boxed{\int w(u)\,dx = \int \left[w(u)\frac{dx}{du}\right] du.}$$

Let's see how this works by applying it to a few problems.

Go to **310**.

310

Consider the problem of evaluating the integral

$$\int xe^{-x^2}\,dx.$$

(continued)

Let $u = x^2$, or $x = \sqrt{u}$. Hence $dx/du = 1/2\sqrt{u} = 1/2x$. Using the rule for change of variable, $\int w(u)\, dx = \int \left[w(u)\frac{dx}{du}\right] du$, the integral becomes

$$\int xe^{-u}\frac{1}{2x}\, du = \frac{1}{2}\int e^{-u}\, du = -\frac{1}{2}e^{-u} + c = -\frac{1}{2}e^{-x^2} + c.$$

To prove that this result is correct, note that

$$\frac{d}{dx}\left(-\frac{1}{2}e^{-x^2} + c\right) = xe^{-x^2},$$

as required.

Try the following somewhat tricky problem. If you need a hint, see frame **311**.

Evaluate $I = \int \sin\theta \cos\theta\, d\theta$.

To check your answer, go to **311**.

311

Let $u = \sin\theta$. Then $\frac{du}{d\theta} = \cos\theta$, and by the rule for change of variable,

$$\int \sin\theta\cos\theta\, d\theta = \int u\cos\theta\, \frac{1}{\cos\theta}\, du$$

$$= \int u\, du = \frac{1}{2}u^2 + c = \frac{1}{2}\sin^2\theta + c.$$

Go to **312**.

312

Here is an example of a simple change of variable. The problem is to calculate $\int \sin 3x\, dx$. If we let $u = 3x$, then the integral is $\sin u$, which is easy to integrate. Using $dx = du/3$, we have

$$\int \sin 3x\, dx = \frac{1}{3}\int \sin u\, du = \frac{1}{3}(-\cos u + c)$$

$$= \frac{1}{3}(\cos 3x + c).$$

To see whether you have caught on, evaluate

$$\int \sin \frac{x}{2} \cos \frac{x}{2}\, dx.$$

(You may find the integral table in frame **306** helpful.)

$$\int \sin \frac{x}{2} \cos \frac{x}{2}\, dx = \underline{\qquad\qquad}.$$

To check your answer, go to **313**.

313

$$\int \sin \frac{x}{2} \cos \frac{x}{2}\, dx = \sin^2 \frac{x}{2} + c.$$

If you obtained this result, go right on to **314**. Otherwise, continue here. If we let $u = x/2$, then $dx = 2\, du$ and

$$\int \sin \frac{x}{2} \cos \frac{x}{2}\, dx = 2 \int \sin u \cos u\, du.$$

From formula 15 of frame **306** we have

$$\int \sin u \cos u\, du = \frac{1}{2} \sin^2 u + c = \frac{1}{2} \sin^2 \frac{x}{2} + c,$$

so

$$\int \sin \frac{x}{2} \cos \frac{x}{2}\, dx = 2\left(\frac{1}{2} \sin^2 \frac{x}{2} + c\right) = \sin^2 \frac{x}{2} + C.$$

($C = 2c =$ any constant.)

Let's check this result:

$$\frac{d}{dx}\left(\sin^2 \frac{x}{2} + C\right) = 2\left(\sin \frac{x}{2} \cos \frac{x}{2}\right)\left(\frac{1}{2}\right) = \sin \frac{x}{2} \cos \frac{x}{2}$$

as required. (We have used the chain rule here.)

Go to **314**.

314

Try to evaluate $\int \frac{dx}{a^2 + b^2x^2}$, where a and b are constants. The integral table in frame **306** may be helpful.

$$\int \frac{dx}{a^2 + b^2x^2} = \underline{\hspace{4cm}}.$$

Go to **315** for the solution.

315

If we let $u = bx$, then $dx = du/b$ and

$$\int \frac{dx}{a^2 + b^2x^2} = \frac{1}{b}\int \frac{du}{a^2 + u^2}$$

$$= \frac{1}{ab}\left(\tan^{-1}\frac{u}{a} + c\right) \qquad \text{(Frame 306, formula 16)}$$

$$= \frac{1}{ab}\left(\tan^{-1}\frac{bx}{a} + c\right).$$

Go to **316**.

316

We have seen how to evaluate an integral by changing the variable from x to $u = ax$, where a is some constant. Often it is possible to simplify an integral by substituting still other quantities for the variable. Here is an example. Evaluate.

$$\int \frac{x\,dx}{x^2 + 4}.$$

Suppose we let $u^2 = x^2 + 4$. Then $2u\,du = 2x\,dx$, and

$$\int \frac{x\,dx}{x^2 + 4} = \int \frac{u\,du}{u^2} = \int \frac{du}{u} = \ln u + c = \ln \sqrt{x^2 + 4} + c.$$

Try to use this method for evaluating the following integral:

$$\int x\sqrt{1 + x^2}\,dx.$$

Answer: ______________

Go to **317** to check your answer.

317

Taking $u^2 = 1 + x^2$, then $2u\,du = 2x\,dx$ and

$$\int x\sqrt{1 + x^2}\,dx = \int u(u\,du) = \int u^2\,du = \frac{1}{3}u^3 + c$$

$$= \frac{1}{3}(1 + x^2)^{3/2} + c.$$

Go to **318**.

318

A technique known as *integration* by *parts* is sometimes helpful. Suppose u and v are any two functions of x. Then, using the product rule for differentiation,

$$\frac{d}{dx}(uv) = u\frac{dv}{dx} + v\frac{du}{dx}.$$

Now integrate both sides of the equation with respect to x.

$$\int \frac{d}{dx}(uv)\,dx = \int u\frac{dv}{dx}\,dx + \int v\frac{du}{dx}\,dx,$$

$$\int d(uv) = \int u\,dv + \int v\,du.$$

But, $\int d(uv) = uv$, and after transposing, we have

$$\int u\,dv = uv - \int v\,du.$$

(continued)

Here is an example: Find $\int \theta \sin \theta \, d\theta$.

Let $u = \theta$, $dv = \sin \theta \, d\theta$. Then it is easy to see that $du = d\theta$, $v = -\cos \theta$. Thus

$$\int \theta \sin \theta \, d\theta = \int u \, dv = uv - \int v \, du$$
$$= -\theta \cos \theta - \int (-\cos \theta) \, d\theta$$
$$= -\theta \cos \theta + \sin \theta.$$

Go to **319**.

319

Try to use integration by parts to find $\int xe^x \, dx$.
Answer (constant omitted):

[$(x - 1)\,e^x$ | xe^x | e^x | $xe^x + x$ | none of these]

If right, go to **321**.
If you missed this, or want to see how to solve the problem, go to **320**.

320

To find $\int xe^x \, dx$ using the formula for integration by parts, we can let $u = x$, $dv = e^x \, dx$, so that $du = dx$, $v = e^x$. Then,

$$\int xe^x \, dx = xe^x - \int e^x \, dx$$
$$= xe^x - e^x = (x - 1)e^x.$$

Go to **321**.

321

Find the following integral using the method of integration by parts:

$$\int x \cos x \, dx.$$

Answer: ________________

Check your answer in **322**.

322

$$\int x \cos x \, dx = x \sin x + \cos x + c.$$

If you want to see the derivation of this, continue here. Otherwise, go on to **323**.

Let us make the following substitution and integrate by parts:

$$u = x, \qquad dv = \cos x \, dx.$$

Thus $du = dx$, $v = \sin x$.

$$\int x \cos x \, dx = \int u \, dv = uv - \int v \, du = x \sin x - \int \sin x \, dx$$
$$= x \sin x + \cos x + c.$$

Go to **323**.

323

In integration problems it is often necessary to use a number of different integration procedures in a single problem.

Try the following (b is a constant):

(a) $\int (\cos 5\theta + b) \, d\theta =$ ________________

(continued)

(b) $\int x \ln x^2 \, dx =$ ______________

Go to **324** for the answers.

324

The correct answers are

(a) $\int(\cos 5\theta + b)\, d\theta = \frac{1}{5} \sin 5\theta + b\theta + c$

(b) $\int x \ln x^2 \, dx = \frac{1}{2} [x^2 (\ln x^2 - 1) + c]$

If you did both of these correctly, you are doing fine—jump ahead to frame **326**. If you missed either problem, go to frame **325**.

325

If you missed (a), you may have been confused by the change in notation from x to θ. Remember x is just a general symbol for a variable. All the integration formulas could be written with θ, or z, or whatever you wish replacing the x. Now for (a) in detail:

$$\begin{aligned} \int (\cos 5\theta + b)\, d\theta &= \int \cos 5\theta \, d\theta + \int b \, d\theta \\ &= \frac{1}{5} \int \cos 5\theta \, d(5\theta) + \int b \, d\theta \\ &= \frac{1}{5} \sin 5\theta + b\theta + c. \end{aligned}$$

For problem (b), let $u = x^2$, $du = 2x \, dx$:

$$\int x \ln x^2 \, dx = \frac{1}{2} \int \ln u \, du = \frac{1}{2} (u \ln u - u + c).$$

[The last step uses formula 9, frame **306**.] Therefore,

$$\int x \ln x^2 \, dx = \frac{1}{2} (x^2 \ln x^2 - x^2 + c).$$

You could also have solved this problem by integration by parts.

Go to **326**.

Answer: (319) $(x - 1)\, e^x$

More on the Area under a Curve

326

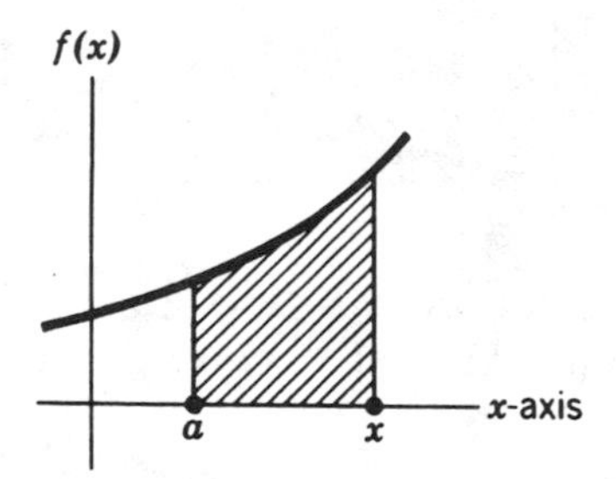

The idea of integration was introduced by the problem of finding the area $A(x)$ under a curve, $f(x)$. In frame **297** we showed that

$$A'(x) = f(x),$$

and we have learned a number of techniques for finding the integral of a function. In general, it can be written

$$A(x) = F(x) + c.$$

$F(x)$ is any particular antiderivative of $f(x)$ and c is an arbitrary constant. However, given $f(x)$ and the interval bounded by a and some value x, there is nothing arbitrary about the area. Thus, to find the area, we must find the correct value for the constant c.

The simplest way to do this is to note that if we take $x = a$, the area has vanishing width and must therefore itself be zero. Hence

$$A(a) = F(a) + c = 0.$$

So, $c = -F(a)$, and the area is given by

$$A(x) = F(x) - F(a).$$

Go to **327**.

327

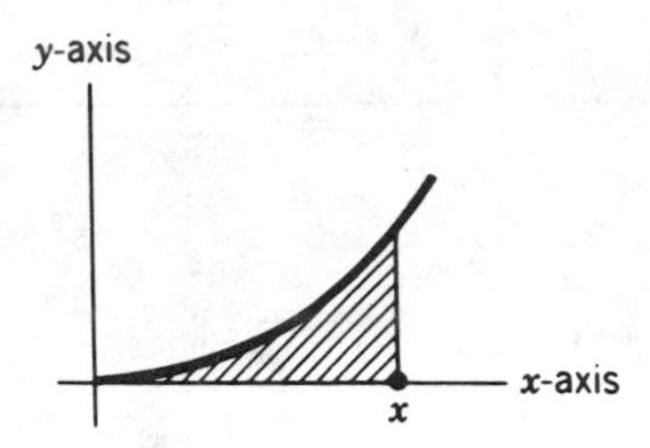

To see how all this works, we will find the area under the curve $y = x^2$ between $x = 0$ and some value of x. Now

$$\int x^2\,dx = \frac{1}{3}x^3 + c = F(x).$$

$$A(x) = F(x) - F(0) = \frac{1}{3}x^3 + c - \left(\frac{1}{3}0^3 + c\right)$$

$$= \frac{1}{3}x^3.$$

Note that the undetermined constant c drops out, as indeed it must. This occurs whenever we evaluate an expression such as $F(x) - F(a)$, so we can simply omit the c. We'll do this in the next few frames.

Go to **328**.

328

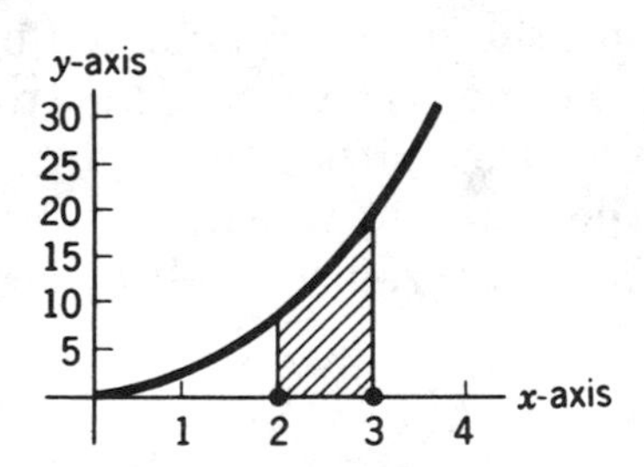

Can you find the area under the curve $y = 2x^2$, between the points $x = 2$ and $x = 3$?

A = [13 | ⅓ | 38⁄3 | 18]

If right, go to **330**.
Otherwise, go to **329**.

329

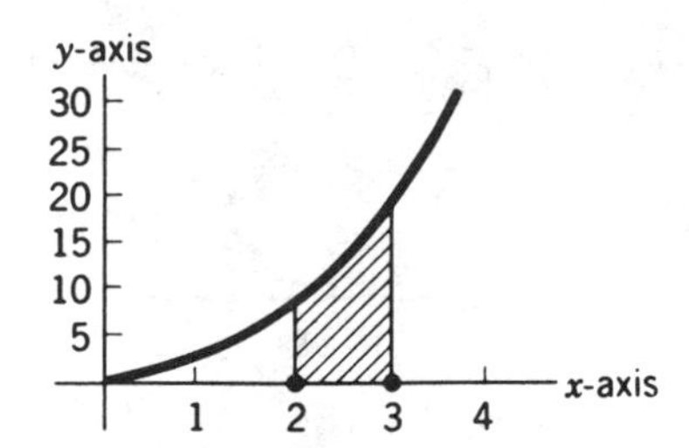

Here is how to solve the problem:

$$A = F(3) - F(2), \qquad F(x) = \int 2x^2\,dx = \frac{2}{3}x^3,$$

$$A = \frac{2}{3} \times 27 - \frac{2}{3} \times 8 = 18 - \frac{16}{3} = \frac{38}{3}.$$

Go to **330**.

330

Find the area under the curve $y = 4x^3$ between $x = -1$ and $x = 2$.

$A =$ [17 | 15/4 | 15 | −16]

Go to **331**.

331

Before we go on, let's introduce a little labor-saving notation.

Frequently we have to find the difference of an expression evaluated at two points, as $F(b) - F(a)$. This is often denoted by

$$F(b) - F(a) = F(x)\Big|_a^b.$$

For instance, $x^2\Big|_a^b = b^2 - a^2$.

(continued)

As another example, the solution to last problem could be written

$$A = \int_{-1}^{2} 4x^3\, dx = x^4 \Big|_{-1}^{2} = 2^4 - (-1)^4 = 16 - 1 = 15.$$

Go to **332**.

332

Let's do one more practice problem:

The graph shows a plot of $y = x^3 + 2$. Find the area between the curve and the x-axis from $x = -1$ to $x = +2$.

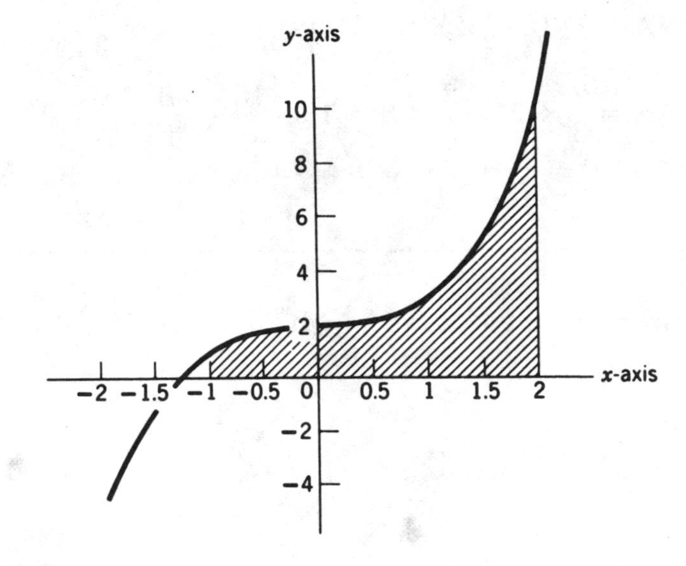

Answer: [5 | ¼ | 4 | 17⁄4 | 39⁄4 | none of these]

If right, go to **334**.
Otherwise, go to **333**.

Answers: **(328)** 38⁄3
(330) 15

333

Here is how to do the problem:

$$A = F(2) - F(-1) = F(x)\Big|_{-1}^{2},$$

$$F = \int y\,dx = \int (x^3 + 2)\,dx = \frac{1}{4}x^4 + 2x,$$

$$A = \left(\frac{1}{4}x^4 + 2x\right)\Big|_{-1}^{2} = \left(\frac{16}{4} + 4\right) - \left(\frac{1}{4} - 2\right) = \frac{39}{4}.$$

Go to **334**.

Definite Integrals

334

In this section we are going to find another way to compute the area under a curve. Our new result will be equivalent to that of the last section, but it will give us a different point of view.

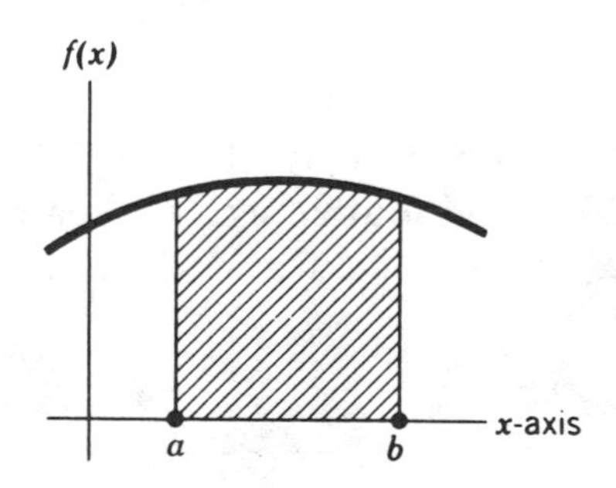

Let's briefly summarize the last section. If A is the area under the curve of $f(x)$ between $x = a$ and some value x, then we showed (frame **297**) that $dA/dx = f(x)$. From this we went on to show (frame **326** that if $F(x)$ is an indefinite integral of $f(x)$, i.e., $F' = f$, then the area under $f(x)$ between the two values of x, a and b, is given by

$$A = F(b) - F(a).$$

Now for a new approach!

Go to **335**.

335

Let's evaluate the area under a curve in the following manner:

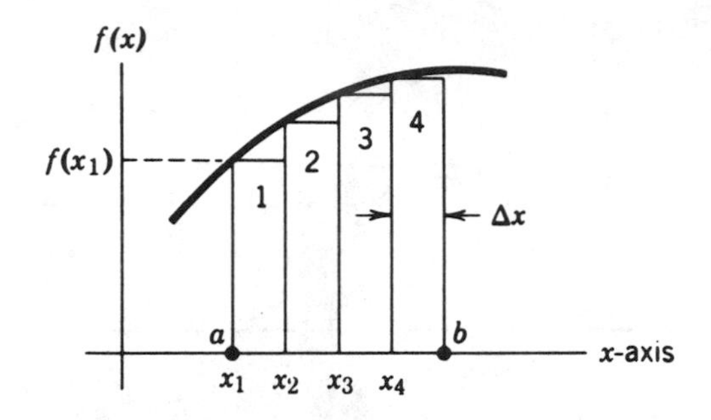

First we divide the area into a number of strips of equal widths by drawing lines parallel to the axis of $f(x)$. The figure shows four such strips drawn. The strips have irregular tops, but we can make them rectangular by drawing a horizontal line at the top of each strip as shown. Suppose we label the strips 1, 2, 3, 4. The width of each strip is

$$\Delta x = \frac{b - a}{4}.$$

The height of the first strip is $f(x_1)$, where x_1 is the value of x at the beginning of the first strip. Similarly, the height of strip 2 is $f(x_2)$, where $x_2 = x_1 + \Delta x$. The third and fourth strips have heights $f(x_3)$ and $f(x_4)$, respectively, where $x_3 = x_1 + 2\ \Delta x$ and $x_4 = x_1 + 3\ \Delta x$.

Go to **336**.

336

You should be able to write an approximate expression for the area of any of the strips. If you need help, review frame **297**. Below write the approximate expression for the area of strip number 3, ΔA_3,

$$\Delta A_3 \approx \underline{\hspace{4cm}}.$$

For the correct answer, go to **337**.

Answer: (332) 39⁄4

337

The approximate area of strip number 3 is $\Delta A_3 \approx f(x_3)\,\Delta x$.

If you want to see a discussion of this, refer again to frame **297**.

Can you write an approximate expression for A, the total area of all four strips?

$$A \approx \underline{\hspace{8cm}}.$$

Try this, and then see **338** for the correct answer.

338

An approximate expression for the total area is simply the sum of the areas of all the strips. In symbols, since $A = \Delta A_1 + \Delta A_2 + \Delta A_3 + \Delta A_4$, we have

$$A \approx f(x_1)\,\Delta x + f(x_2)\,\Delta x + f(x_3)\,\Delta x + f(x_4)\,\Delta x.$$

We could also write this

$$A = \sum_{i=1}^{N} f(x_i)\,\Delta x.$$

$\sum$ is the Greek letter *sigma* which corresponds to the English letter S and stands here for the sum. The symbol $\sum_{i=1}^{N} g(x_i)$ means $g(x_1) + g(x_2) + g(x_3) + \cdots + g(x_N)$.

Go to **339**.

339

Suppose we divide the area into more strips each of which is narrower, as shown in the drawings. Evidently our approximation gets better and better.

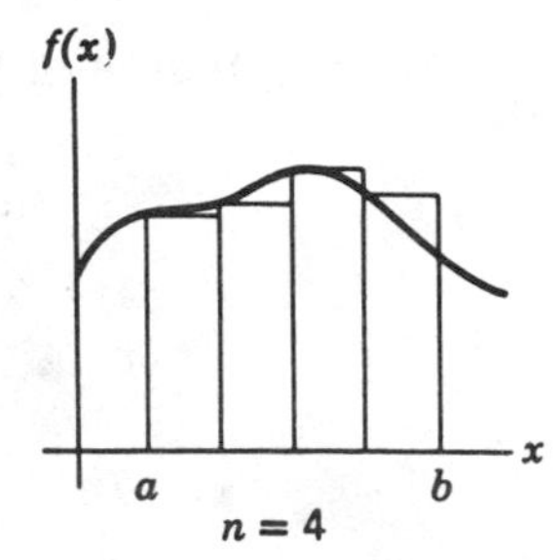

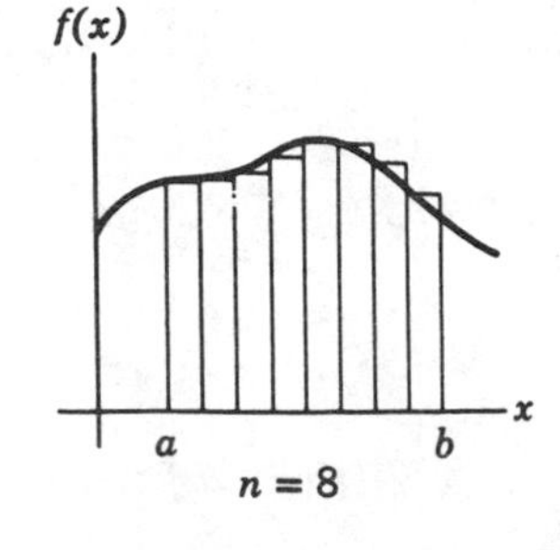

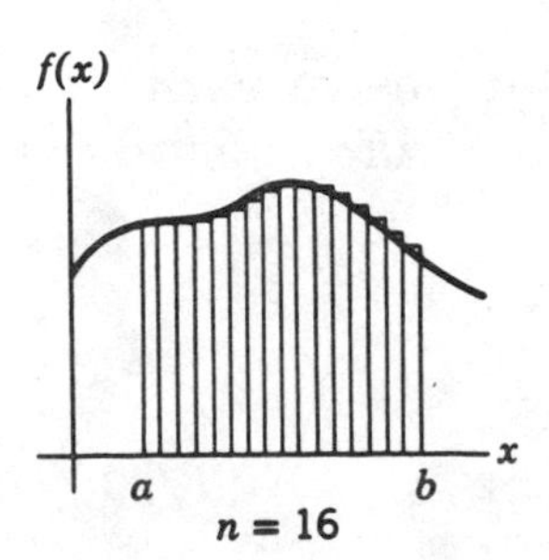

(continued)

If we divide the area into N strips, then $A \approx \sum_{i=1}^{N} f(x_i)\,\Delta x$, where $N = \dfrac{b-a}{\Delta x}$. Now, if we take the limit where $\Delta x \to 0$, the approximation becomes an equality. Thus,

$$A = \lim_{\Delta x \to 0} \sum_{i=1}^{N} f(x_i)\,\Delta x.$$

Such a limit is so important that it is given a special name and symbol. It is called the *definite integral* and is written $\int_a^b f(x)\,dx$. This symbol looks similar to the indefinite integral, $\int f(x)\,dx$, and as we shall see in the next frame, it is related. However, it is important to remember that the definite integral is defined by the limit described above. So, by definition,

$$\boxed{\int_a^b f(x)\,dx = \lim_{\Delta x \to 0} \sum_{i=1}^{N} f(x_i)\,\Delta x.}$$

(Incidentally, the integral symbol $\int$ evolved from the letter S and like sigma it was chosen to stand for *sum*.)

Go to **340**.

340

With this definition for the definite integral, the discussion in the last frame shows that the area A under the curve is equal to the *definite* integral.

$$\boxed{A = \int_a^b f(x)\,dx.}$$

But we saw earlier that the area can also be evaluated in terms of the *indefinite* integral.

$$F(x) = \int f(x)\,dx$$

by

$$A = F(b) - F(a).$$

Therefore we have the general relation

$$\int_a^b f(x)\,dx = F(b) - F(a) = \left\{\int f(x)\,dx\right\}\Big|_a^b$$

Thus the *definite* integral can be expressed in terms of an *indefinite* integral evaluated at the limits. This remarkable result is often called the *fundamental theorem of integral calculus.*

Go to **341**.

341

To help remember the definition of definite integral, try writing it yourself. Write an expression defining the definite integral of $f(x)$ between limits a and b.

To check your answer, go to **342**.

342

The correct answer is

$$\int_a^b f(x)\,dx = \lim_{\Delta x \to 0} \sum_{i=1}^{N} f(x_i)\,\Delta x, \qquad \text{where } N = \frac{b-a}{\Delta x}.$$

Congratulations if you wrote this or an equivalent expression.

If you wrote

$$\int_a^b f(x)\,dx = F(b) - F(a), \qquad \text{where } F(x) = \int f(x)\,dx,$$

your statement is true, but it is not the *definition* of a definite integral. The result is true because both sides represent the same thing—the area under the curve of $f(x)$ between $x = a$ and $x = b$. It is an important result, since without it we would have no way of evaluating the definite integral, but it is not true by definition.

(continued)

If this reasoning is clear to you, go right on to **343**. Otherwise, review the material in this chapter, and then, to see a further discussion of definite and indefinite integrals,

Go on to **343**.

343

Perhaps the definite integral seems an unnecessary complication to you. After all, the only thing we accomplished with it was to write the area under a curve a second way. To actually compute the area, we were led back to the indefinite integral. However, we could have found the area directly from the indefinite integral in the first place. The importance of the definite integral arises from its definition as the *limit* of a sum. The process of dividing a system into little bits and then adding them all together is applicable to many problems. This naturally leads to definite integrals which we can evaluate in terms of indefinite integrals by using the fundamental theorem in frame **340**.

Go to **344**.

344

Can you prove that

$$\int_a^b f(x)\,dx = -\int_b^a f(x)\,dx?$$

After you have tried to prove this result, go to **345**.

345

The proof that $\int_a^b f(x)\,dx = -\int_b^a f(x)\,dx$ is simple.

$$\int_a^b f(x)\,dx = F(b) - F(a), \qquad \text{where } F(x) = \int f(x)\,dx,$$

but

$$\int_b^a f(x)\,dx = F(a) - F(b) = -[F(b) - F(a)]$$
$$= -\int_a^b f(x)\,dx.$$

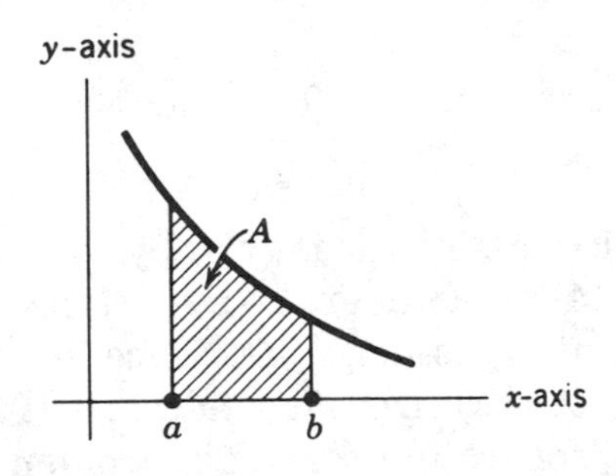

The points a and b are called the limits of the integral [nothing to do with $\lim_{x\to a} f(x)$; here limit simply means the boundary]. The process of evaluating

$$\int_a^b f(x)\,dx$$

is often spoken of as "integrating $f(x)$ from a to b," and the expression is called the "integral of $f(x)$ from a to b."

Go to **346**.

346

Which of the following expressions correctly gives $\int_0^{2\pi} \sin\theta\,d\theta$?

[1 | 0 | 2π | -2 | -2π | none of these]

Go to **347**.

347

$$\int_0^{2\pi} \sin\theta\, d\theta = -\cos\theta \Big|_0^{2\pi} = -(1-1) = 0.$$

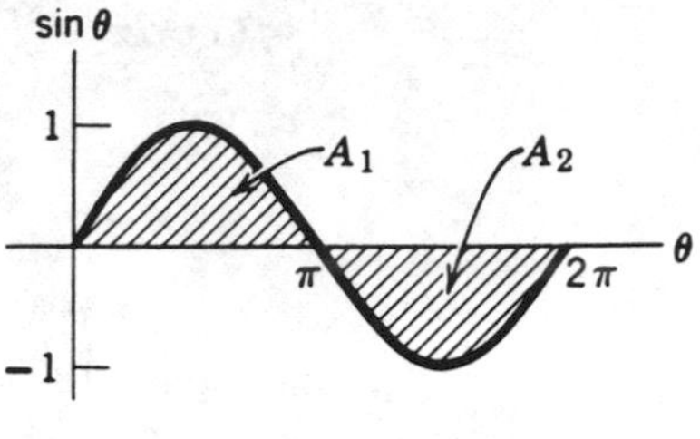

It is easy to see why this result is true by inspecting the figure. The integral yields the total area under the curve, from 0 to 2π, which is the sum of A_1 and A_2. But A_2 is negative, since $\sin\theta$ is negative in that region. By symmetry, the two areas just add to 0. However, you should be able to find A_1 or A_2 separately. Try this problem:

$$A_1 = \int_0^{\pi} \sin\theta\, d\theta = [1 \mid 2 \mid -1 \mid -2 \mid \pi \mid 0]$$

If right, go to **349**.
Otherwise, go to **348**.

348

$$A_1 = \int \sin\theta\, d\theta = -\cos\theta \Big|_0^{\pi} = -[-1-(+1)] = 2.$$

If you forgot the integral, you can find it in the table on page 256. In evaluating $\cos\theta$ at the limits, we need to know that $\cos\pi = -1$, $\cos 0 = 1$.

Go to **349**.

Answer: (346) 0

349

Here is a graph of the function $y = 1 - e^{-x}$.

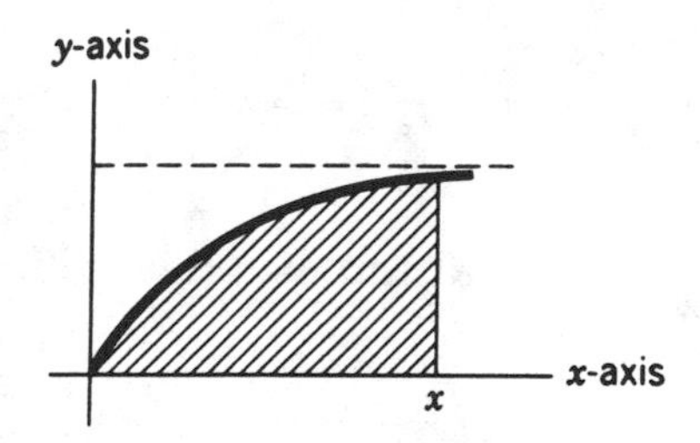

Can you find the shaded area under the curve between the origin and x?

Answer: [e^{-x} | $1 - e^{-x}$ | $x + e^{-x}$ | $x + e^{-x} - 1$]

Go to **351** if you did this correctly.
See **350** for the solution, or if you want to see a discussion of the meaning of the area.

350

Here is the solution to **349**.

$$A = \int_0^x y\,dx = \int_0^x (1 - e^{-x})\,dx = \int_0^x dx - \int_0^x e^{-x}\,dx$$
$$= [x - (-e^{-x})]\Big|_0^x = [x + e^{-x}]\Big|_0^x = x + e^{-x} - 1,$$

The area found is bounded by a vertical line through x. Our result gives A as a variable that depends on x. If we choose a specific value for x, we can substitute it into the above formula for A and obtain a specific value for A. We have evaluated a definite integral in which one of the boundary points has been left as a variable.

Go to **351**.

351

Let's evaluate one more definite integral before going on. Find:

$$\int_0^1 \frac{dx}{\sqrt{1 - x^2}}$$

(If you need to, use the integral tables, page 256.)

Answer: [0 | 1 | ∞ | π | $\pi/2$ | none of these]

If you got the right answer, go to **353**.
If you got the wrong answer, or no answer at all, go to **352**.

352

From the integral table, page 256, we see that

$$\int \frac{dx}{\sqrt{1 - x^2}} = \sin^{-1} x + c.$$

Therefore,

$$\int_0^1 \frac{dx}{\sqrt{1 - x^2}} = \sin^{-1} \Big|_0^1 = \sin^{-1} 1 - \sin^{-1} 0.$$

But $\sin^{-1} 1 = \frac{\pi}{2}$, since $\sin \frac{\pi}{2} = 1$. Similarly, $\sin^{-1} 0 = 0$. Thus, the integral has the value $\frac{\pi}{2} - 0 = \frac{\pi}{2}$.

Answers: **(347)** 2 **(349)** $x + e^{-x} - 1$

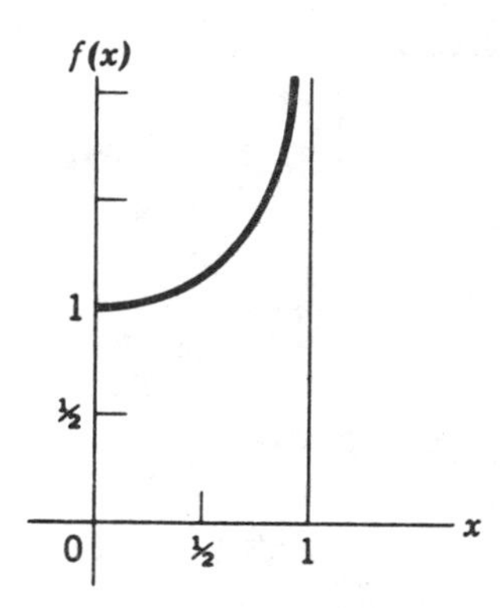

A graph of $f(x) = \dfrac{1}{\sqrt{1 - x^2}}$ is shown above. Although the function is discontinuous at $x = 1$, the area under the curve is perfectly well defined.

Go to **353**.

Numerical Integration

353

In the preceding frames we learned a few of the techniques for integration; the references listed in Appendix B describe many others. There is, however, no general method for finding the indefinite integral of a function. Indefinite integrals of hundreds of functions are known and listed in the integral tables. Many functions can be integrated by some clever change of variable which transforms them into one of the tabulated forms, but integrals for many other functions are simply not known. Nevertheless, definite integrals can always be evaluated numerically. With a computer, numerical integration is often so accurate and efficient that a definite integral can be calculated as easily as if it were tabulated. A computer is not essential, but you will find that a calculator (particularly a programmable calculator) is an enormous time saver in carrying out numerical integration.

In this section we shall describe numerical integration. The one problem has been designed so that it can be worked without a calculator.

Go to **354**.

354

Recall from frame **339** that the definite integral is the limit of a sum

$$\int_a^b y\,dx = \lim_{\Delta \to 0} \sum_{i=1}^{N} y(x_i)\Delta,$$

where $\Delta = (x_b - x_a)/N$. As N increases, the area under the rectangles approaches the area under the curve.

For a finite value of N, the area under the rectangle is not identical to the integral [unless $y(x)$ = constant], but it can be close. This is the basic idea of numerical integration. Here is the procedure.

1. Divide the interval $b - a$ into some convenient number N of equal intervals, $\Delta=(b - a)/N$.
2. Evaluate $y_i = y(x_i)$ at each interval, where $i = 1, 2, \ldots, N$.
3. Multiply each y_i by Δ.
4. Add the results.

The final result is in an approximation to the integral. How good the approximation is depends on the choice of N and the precise method by which the sum is evaluated.

In carrying out the above steps, it may have already occurred to you that a great deal of multiplication is avoided if one first adds all the y_i's and then multiplies the final result by Δ. Thus,

$$S = \sum_{i=1}^{N} (y_i\Delta) = \Delta \sum_{i=1}^{N} y_i$$

Go to **355**.

355

In evaluating the integral numerically, one could choose for y_i the value of y at either end of the interval, as in the drawings. For the function shown, it is evident that one choice underestimates the integral and the other overestimates it. Neither looks particularly accurate. Taking for y_i the value y at either end of the interval is clearly less good than taking it at the midpoint. However, an even better procedure would be to take a suitable weighted average of y at the ends and the middle.

Answer: (351) $\pi/2$

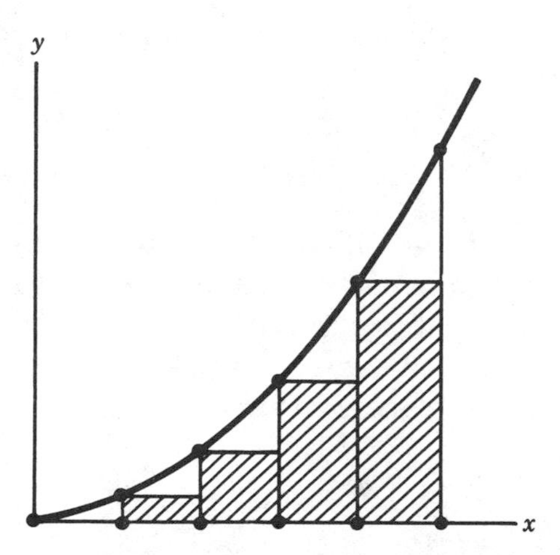

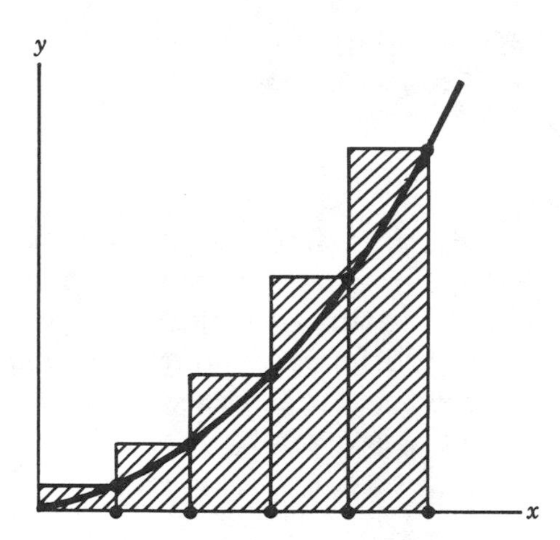

An averaging process that is simple, accurate, and widely used considers the interval in pairs and weights the midpoint of each pair four times that of each end. In that case,

$$\bar{y}_i = \frac{1}{6}(y_{i-1} + 4y_i + y_{i+1}).$$

(Note that the width of this pair of segments is 2Δ, not Δ.) If then the entire interval is divided into an even number of intervals,

$$\int_A^B y\,dx = \frac{2\Delta}{6}(y_0 + 4y_1 + y_2 + 4y_3 + y_4 + \cdots + y_{N-2} + 4y_{N-1} + y_N)$$
$$= \frac{\Delta}{3}(y_0 + 4y_1 + 2y_2 + 4y_3 + 2y_4 + \cdots + 2y_{N-2} + 4y_{N-1} + y_N).$$

This method is called *Simpson's rule.* If you would like to know just why it works so well, go to **356**. Otherwise,

Skip on to **357**.

356

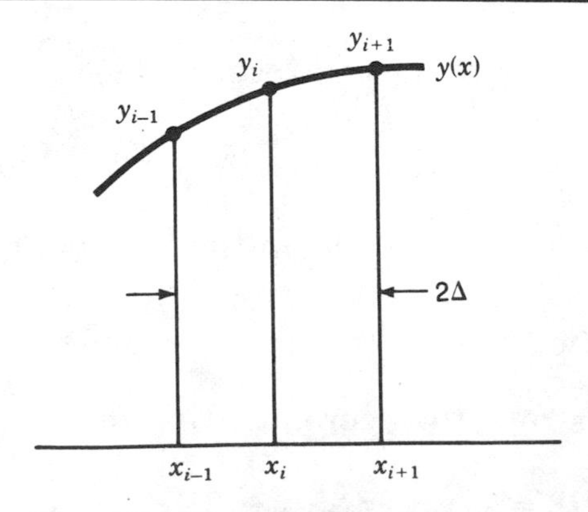

Simpson's rule is based on the idea that the simplest curve which can

(continued)

go through three arbitrary points is a parabola. So, if we assume that in the interval shown

$$y = ax^2 + bx + c$$

and demand that $y_j = ax_j^2 + bx_j + c$, where $j = i-1$, i, and $i+1$ in turn, one has three equations; these are just the number of equations needed to define the three constants, a, b, and c. The algebra of solving for a, b, and c is straightforward but a little lengthy. You may want to work it out for yourself, but in any case, the result is that the area in the segment $x_{i-1} < x < x_{i+1}$ is given by

$$\frac{2\Delta}{6}(y_{i-1} + 4y_i + y_{i+1}),$$

which is the expression used in Simpson's rule.

Go to **357**.

357

Here is an example of how Simpson's rule works. You will not need a calculator.

The goal is to calculate $I = \int_0^{10} x^3\,dx$. We can do this integral exactly, which will make it easy to check the accuracy of the numerical calculation.

$$I = \int_0^{10} x^3\,dx = \frac{1}{4}x^4\Big|_0^{10} = \frac{1}{4} \times 10{,}000 = 2500.$$

We shall take $N = 10$. Then, $\Delta = \frac{10-0}{10} = 1$. $x_0 = 0$, $x_{10} = 10$, and in general, $x_i = (i)(\Delta) = i$.

If we denote the sum of the odd terms by

$$S_{\text{odd}} = S_1 + S_3 + S_5 + S_7 + S_9$$

and the sum of the even term within the interval by

$$S_{\text{even}} = S_2 + S_4 + S_6 + S_8,$$

then, by Simpson's rule, the approximation to the integral is

$$I' = \frac{\Delta}{3}(y_0 + 4S_{\text{odd}} + 2S_{\text{even}} + y_{10}).$$

I' can be calculated using the tables below.

i	x_i	x_i^2	x_i^3
1			
3			
5			
7			
9			
		S_{odd} =	

i	x_i	x_i^2	x_i^3
2			
4			
6			
8			
		S_{even} =	

y_0	
$4S_{odd}$	
$2S_{even}$	
y_{10}	
Sum =	

Then $I' = \frac{\Delta}{3} \times \text{Sum} =$ ____________.

Go to **358**.

358

The answer is $I' = 2501\frac{1}{3}$.

This result is close to the exact value of the integral, $I = 2500$. Considering the relatively small number of points used, this is remarkably accurate.

An interesting exercise in numerical integration is to evaluate π using the relation

$$\tan^{-1} A = \int_0^A \frac{dx}{1 + x^2}$$

(continued)

which follows from formula 16 in Table 2. Because $\pi/4 = \tan^{-1} 1$, one has

$$\pi = 4 \int_0^1 \frac{dx}{1 + x^2}.$$

You may wish to try your skill by integrating other functions whose integrals you know, for instance, $\sin \theta$ or e^{-x}.

It is evident that by numerical integration you can find the definite integral of *any* function, and therein lies its power. With computers it is possible to integrate numerically at very high speed. One must have some criterion for choosing the interval size and be able to deal with problems such as singularities in the integral. Nevertheless, with the simple method described here you can often do surprisingly well.

Go to **359**.

Some Applications of Integration

359

In this section we are going to apply integration to a few simple problems.

In Chapter 2 we learned how to find the velocity of a particle if we know its position in terms of time. Now we can reverse the procedure and find the position from the velocity. For instance, we are in an automobile driving along a straight road through thick fog. To make matters worse, our mileage indicator is broken. Instead of watching the road all the time, let's keep an eye on the speedometer. We have a good watch along, and we make a continuous record of the speed starting from the time when we were at rest. The problem is to find how far we have gone. (This is a dangerous method for navigating a car, but it is actually used for navigating submarines and spacecraft.) More specifically, given $v(t)$, how do we find $S(t)$, the distance traveled since time t_0 when we were at rest? Try to work out a method.

$S(t) =$ ______________

To check your result, go to **360**.

360

Since

$$v = \frac{dS}{dt},$$

we must have $dS = v\,dt$ (as was shown in **263**).

Now let us integrate both sides from the initial point ($t = t_0$, $S = 0$) to the final point (t, S).

We have

$$\int_{S=0}^{S} dS = \int_{t_0}^{t} v\,dt,$$

so

$$S = \int_{t_0}^{t} v\,dt.$$

If you did not get this result, or would like to see more explanation, go to **361**.
Otherwise, go to **362**.

361

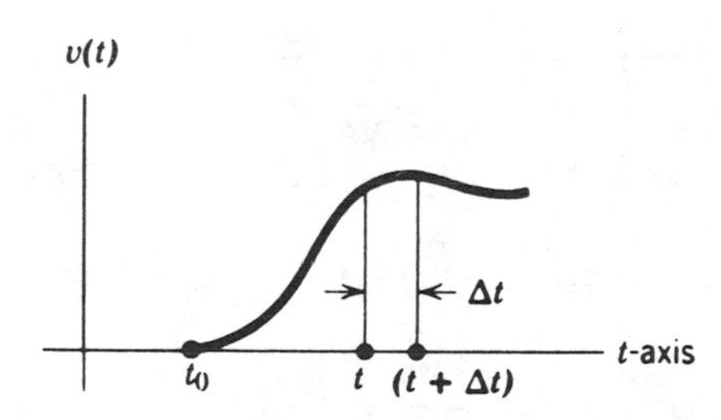

Another way to understand this problem is to look at it graphically. Here is a plot of $v(t)$ as a function of t. In time Δt the distance traveled is $\Delta S = v\,\Delta t$. The total distance traveled is thus equal to the area under the curve between the initial time and the time of interest, and this is $\int_{t_0}^{t} v(t)\,dt$.

Go to **363**.

362

Suppose an object moves with a velocity which continually decreases in the following way.

$$v(t) = v_0 e^{-bt}$$

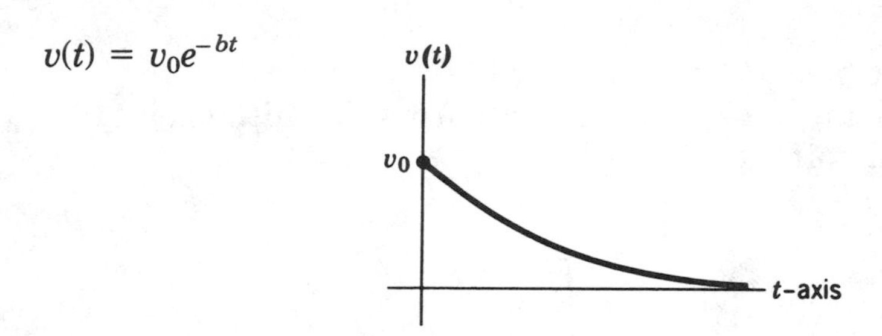

(v_0 and b are constants). At $t = 0$ the object is at the origin; $S = 0$. Which of the following is the distance the object will have moved after an infinite time (or, if you prefer, after a very long time)?

$$[0 \mid v_0 \mid v_0 e^{-1} \mid \frac{v_0}{b} \mid \infty]$$

If correct, go to **364**.
Otherwise, go to **363**.

363

Here is the solution to the problem of frame **362**.

$$S(t) - S(0) = \int_0^t v\, dt = \int_0^t v_0\, e^{-bt}\, dt$$

$$S(t) - 0 = -\frac{v_0}{b}\, e^{-bt}\Big|_0^t = -\frac{v_0}{b}\,(e^{-bt} - 1).$$

We are interested in $\lim_{t\to\infty} S(t)$, but since $e^{-bt} \to 0$ as $t \to \infty$, we have

$$\lim_{t\to\infty} S(t) = -\frac{v_0}{b}\,(0 - 1) = \frac{v_0}{b}.$$

Although the object never comes completely to rest, its velocity gets so small that the total distance traveled is finite.

Go to **364**.

364

Not all integrals give finite results. For example, try this problem.

A particle starts from the origin at $t = 0$ with a velocity $v(t) = v_0/(b + t)$, where v_0 and b are constants.

How far does it travel as $t \to \infty$?

$$\left[v_0 \ln \frac{1}{b} \mid \frac{v_0}{b} \mid \frac{v_0}{b^2} \mid \text{none of these}\right]$$

Go to **365**.

365

It is easy to see that problem **364** leads to an infinite integral.

$$S(t) - 0 = \int_{t=0}^{t} v_0 \frac{dt}{b + t} = v_0 \ln(b + t)\Big|_0^t$$

$$= v_0 [\ln(b + t) - \ln b]$$

$$= v_0 \ln\left(1 + \frac{t}{b}\right)$$

Since $\ln(1 + t/b) \to \infty$ as $t \to \infty$, we see that $S(t) \to \infty$ as $t \to \infty$.

In this case, the particle is always moving fast enough so that its motion is unlimited. Or, alternatively, the area under the curve $v(t) = v_0/(b + t)$ increases without limit as $t \to \infty$.

Go to **366**.

366

Integration can be used for many tasks besides calculating the area under a curve. For example, it can be used to find the volumes of solids of known geometry. A general method for this is explained in frame **380**. However, one can calculate the volume of symmetric solids by a simple

(continued)

extension of methods we have learned already. In the next few frames we are going to find the volume of a right circular cone.

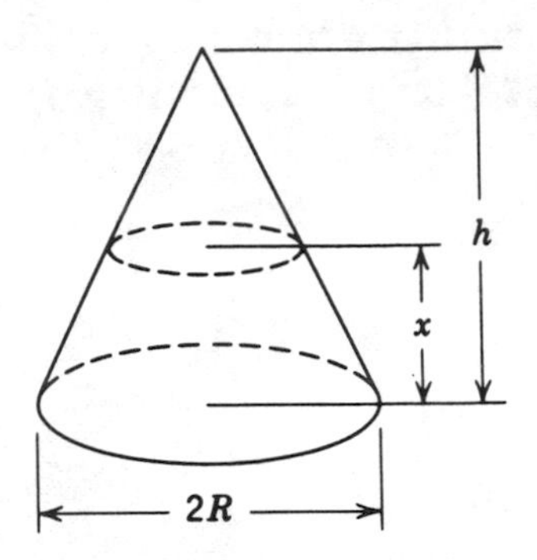

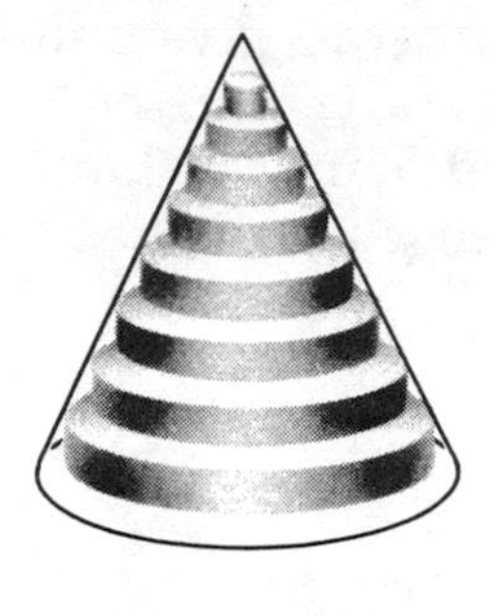

The height of the cone is h, and the radius of the base is R. We will let x represent distance vertically from the base.

Our method of attack is similar to that used in frame **339** to find the area under a curve. We will slice the body into a number of discs whose volume is approximately that of the cone in the figure (the cone has been approximated by eight circular discs). Then we have

$$V \cong \sum_{i=1}^{8} \Delta V_i,$$

where ΔV_i is the volume of one of the discs. In the limit where the height of each disc (and hence the volume) goes to 0, we have

$$V = \int dV.$$

In order to evaluate this, we have to have an expression for dV. To find this,

Go to frame **367**.

367

Because we are going to take the limit where $\Delta V \to 0$, we will represent the volume element by dV from the start.

Below is a picture of a section of the cone, which for our purposes is represented by a disc. The radius of the disc is r and its height is dx. Try to find an expression for dV in terms of x. (You will have to find r in terms of x).

Answer: (363) $\dfrac{v_0}{b}$

(364) none of these

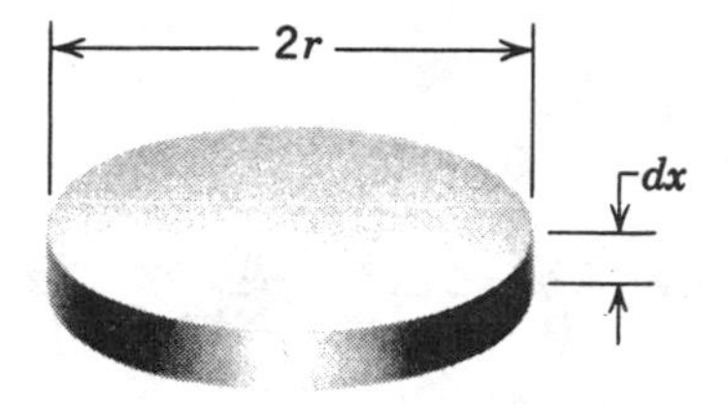

$$dV = \_\_\_\_\_\_\_\_\_\_\_\_.$$

To check your result, or to see how to obtain the result, go to **368**.

368

$$dV = \pi R^2\left(1 - \frac{x}{h}\right)^2 dx.$$

If you got this answer, go on to **369**.

If you want to see how to derive it, read on.

The volume of this disc is the product of the area and height. Thus, $dV = \pi r^2\, dx$. Our remaining task is to express r in terms of x.

The diagram shows a cross section of the cone. Since r and R are corresponding edges of similar triangles, it should be clear that

$\frac{r}{R} = \frac{h - x}{h}$, or $r = R\left(1 - \frac{x}{h}\right)$. Thus,

$$dV = \pi r^2\, dx = \pi R^2\left(1 - \frac{x}{h}\right)^2 dx.$$

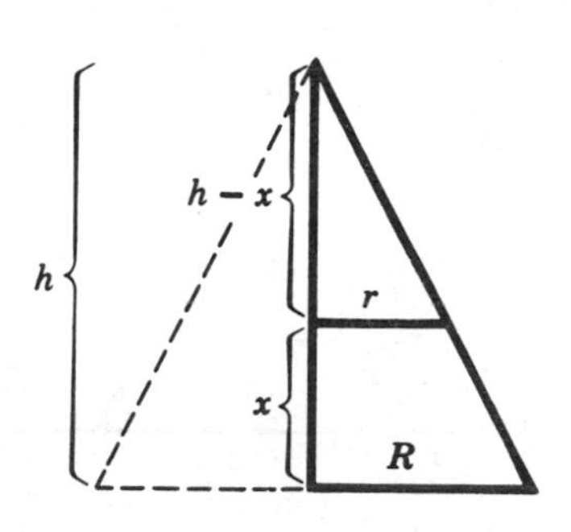

Go to **369**.

369

We now have an integral for V.

$$V = \int_0^h dV = \int_0^h \pi R^2 \left(1 - \frac{x}{h}\right)^2 dx.$$

Try to evaluate this.

$V =$ ______________.

To check your answer, go to **370**.

370

You should have obtained the result

$$V = \frac{1}{3} \pi R^2 h.$$

Congratulations, if you did. Go on to **371**. Otherwise, read below:

$$V = \int_0^h \pi R^2 \left(1 - \frac{x}{h}\right)^2 dx = \pi R^2 \int_0^h \left(1 - \frac{2x}{h} + \frac{x^2}{h^2}\right) dx$$

$$= \pi R^2 \left(x - \frac{x^2}{h} + \frac{1}{3}\frac{x^3}{h^2}\right)\Big|_0^h = \pi R^2 \left(h - h + \frac{1}{3}h\right)$$

$$= \frac{1}{3} \pi R^2 h.$$

Go to **371**.

371

Here is one more problem. Let's find the volume of a sphere.

It will simplify matters if we find the volume of a hemisphere, V', which is just half the required volume, V. Thus, $V' = V/2$.

Can you write an integral which will give the volume of the hemisphere? (The slice of the hemisphere shown in the drawing may help you in this.)

$V' =$ ______________.

Go to **372** to check your formula.

372

The answer is

$$V' = \int_0^R \pi(R^2 - x^2)\, dx.$$

If you wrote this, go ahead to frame **373**. Otherwise, read on.

(continued)

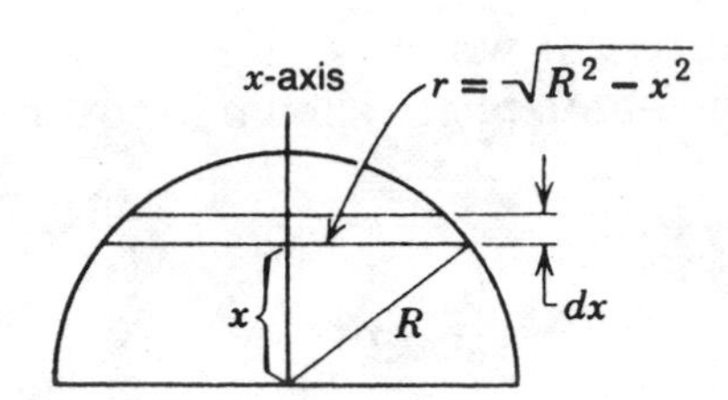

Here is a vertical section through the hemisphere. The volume of the disc between x and $x + dx$ is $\pi r^2\,dx$. But, as can be seen from the triangle indicated, $x^2 + r^2 = R^2$, so

$$r^2 = R^2 - x^2.$$

Hence, $dV' = \pi(R^2 - x^2)\,dx$ and $V' = \int_0^R \pi(R^2 - x^2)\,dx$.

Go to **373**.

373

Now go ahead and evaluate the integral

$$V' = \int_0^R \pi(R^2 - x^2)\,dx.$$

$V' =$ ____________________.

To see the correct answer, go to **374**.

374

$$V' = \int_0^R \pi(R^2 - x^2)\,dx = \pi\left(R^2x - \frac{1}{3}x^3\right)\Big|_0^R$$

$$= \pi\left(R^3 - \frac{1}{3}R^3\right) = \frac{2}{3}\pi R^3.$$

Since $V = 2V'$, $V = \frac{4}{3}\pi R^3$.

Go to **375**.

Multiple Integrals

375

Although the subject of this section—multiple integrals—is essential for some problems, it is not needed for many others. Multiple integrals are also a little complicated. Therefore, if you feel you have had about as much calculus as you want right now, you should skip to the conclusion, frame **384**.

The integrals we have discussed so far, of the form $\int f(x)\,dx$, have had a single independent variable, usually called x. Double integrals are similarly defined for *two* independent variables, x and y. In general, multiple integrals are defined for an arbitrary number of independent variables, but we will only consider two. Note that up to now y has often been the dependent variable: $y = f(x)$. In this section, however, y along with x will always be an independent variable and $z = f(x, y)$ will be the dependent variable. Thus, z is a function of two variables.

In frame **339** the definite integral of $f(x)$ between a and b was defined by

$$\int_a^b f(x)\,dx = \lim_{\Delta x \to 0} \sum_{i=1}^{N} f(x_i)\,\Delta x.$$

The double integral is similarly defined, but with two independent variables. There are, however, some important differences. For a single definite integral the integration takes place over a closed interval between a and b on the x axis. In contrast, the integration of $f(x, y)$ takes place over a closed region R in the x-y plane.

To define the double integral, divide the region R into N smaller regions each of area ΔA_i.

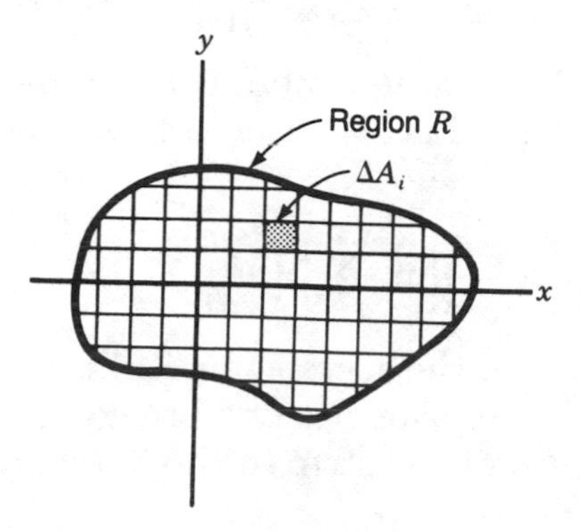

(continued)

Let x_i, y_i be an arbitrary point inside the region ΔA_i. Then in analogy to the integral of a single variable, the double integral is defined as

$$\iint f(x, y)\, dA = \lim_{\Delta A_i \to 0} \sum_{i=1}^{N} f(x_i, y_i)\, \Delta A_i.$$

Go to **376**.

376

The double integral is often evaluated by taking ΔA_i to be a small rectangle with sides parallel to the x and y axes, and by first evaluating the sum and limit along one direction and then along the other. Consider the upper portion of the region R in the x-y plane to be bounded by the curve $y = g_2(x)$, while the lower portion is bounded by $g_1(x)$, as in the diagram.

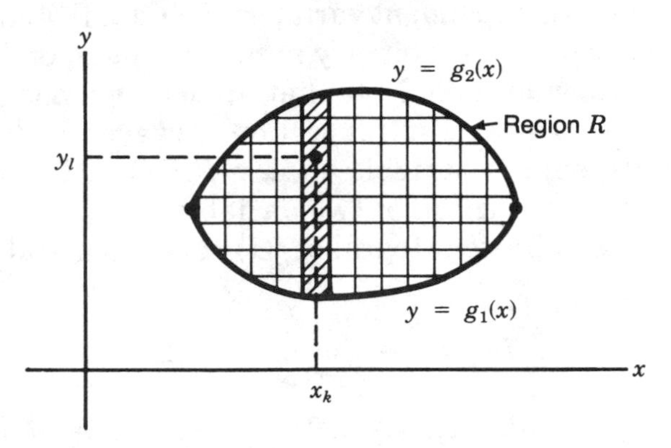

If we let $\Delta A_i = \Delta x_k\, \Delta y_l$, then

$$\iint_R f(x, y)\, dA = \lim_{\Delta A_i \to 0} \sum_{i=1}^{N} f(x_i, y_i)\, \Delta A_i$$

$$= \lim_{\Delta x_k \to 0} \lim_{\Delta y_l \to 0} \sum_{k=1}^{p} \sum_{l=1}^{q} f(x_k, y_l)\, \Delta y_l\, \Delta x_k.$$

This is a complicated expression, but it can be simplified by carrying it out in two separate steps. Let us insert some brackets to clarify the separate steps.

$$\iint_R f(x, y)\, dA = \lim_{\Delta x_k \to 0} \sum_{k=1}^{k} \left[\lim_{\Delta y_l \to 0} \sum_{l=1}^{q} f(x_k, y_l)\, \Delta y_l \right] \Delta x_k.$$

The first step is to carry out the operation within the brackets. Note that x_k is not altered as we sum over l in the brackets. This corresponds to summing over the crosshatched strip in the diagram with x_k treated as approximately a constant. The quantity in square brackets is then merely a definite integral of the variable y, with x treated as a constant. Note that although the limits of integration, $g_1(x)$ and $g_2(x)$, are constants for a particular value of x, they are in general nonconstant functions of x. The quantity in square brackets can then be written as

$$\int_{g_1(x_k)}^{g_2(x_k)} f(x_k, y)\, dy.$$

This quantity will no longer depend on y, but it will depend on x_k both through the integrand $f(x_k, y)$ and the limits $g_1(x_k), g_2(x_k)$. Consequently,

$$\iint_R f(x, y)\, dA = \lim_{\Delta x_k \to 0} \sum_{k=1}^{k} \left[\int_{g_1(x_k)}^{g_2(x_k)} f(x_k, y)\, dy\right] \Delta x_k$$

$$= \int_a^b \left[\int_{g_1(x)}^{g_2(x)} f(x, y)\, dy\right] dx.$$

In calculations it is essential that one first evaluate the integral in the square brackets while treating x as a constant. The result is some function which depends only on x. The next step is to calculate the integral of this function with respect to x, treating x now as a variable.

The double integral expressed in the above form is often called the *iterated integral.*

Go to **377**.

377

Multiple integrals are most easily evaluated if the region R is a rectangle whose sides are parallel to the x and y coordinate axes, as shown in the drawing.

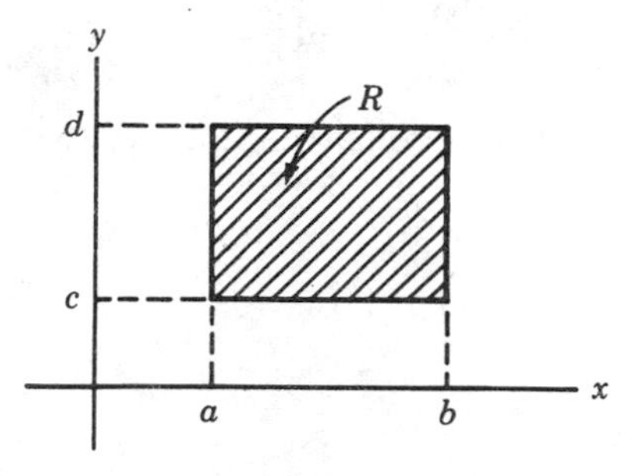

The double integral is

$$\iint_R f(x, y)\, dA = \int_a^b \left[\int_c^d f(x, y)\, dy\right] dx.$$

(continued)

As an exercise to test your understanding, how would the above expression be written if the integration over x were to be carried over before the integration over y.

Go to **378**.

378

The double integral, integrated first over x, is written

$$\iint_R f(x, y)\, dA = \int_c^d \left[\int_a^b f(x, y)\, dx \right] dy.$$

This can be found merely by interchanging the y and x operations in the evolution of double integrals. (Note that the integration limits have to be interchanged at the same time.)

To see how this works, let us evaluate the double integral of $f(x, y) = 3x^2 + 2y$ over the rectangle in the x-y plane bounded by the lines $x = 0$, $x = 3$, $y = 2$, and $y = 4$.

The double integral is equal to the iterated integral.

$$\iint_R \left(\frac{1}{3}x^2 + y\right) dA = \int_0^3 \left[\int_2^4 (3x^2 + 2y)\, dy \right] dx.$$

Alternatively we could have written

$$\iint_R \left(\frac{1}{3}x^2 + y\right) dA = \int_2^4 \left[\int_0^3 (3x^2 + 2y)\, dy \right] dx.$$

Evaluate each of the above expressions. The answers should be the same.

Integral = ______________.

If you made an error or want more explanation, go to **379**.
Otherwise, go to **380**.

379

By the first expression,

$$\int_0^3 \left[\int_2^4 (3x^2 + 2y)\, dy\right] dx = \int_0^3 (3x^2y + y^2)\Big|_2^4 dx$$

$$= \int_0^3 [3x^2(4-2) + (16-4)]\, dx = \int_0^3 (6x^2 + 12)\, dx$$

$$= \left(6\frac{x^3}{3} + 12x\right)\Big|_0^3 = 54 + 36 = 90.$$

By the second expression,

$$\int_2^4 \left[\int_0^3 (3x^2 + 2y)\, dx\right] dy = \int_2^4 (x^3 + 2yx)\Big|_0^3 dy$$

$$= \int_2^4 (27 + 6y)\, dy = (27y + 3y^2)\Big|_2^4$$

$$= 108 + 48 - (54 + 12) = 90.$$

Go to **380**.

380

Just as the equation $y = f(x)$ defines a curve in the two dimensional x-y plane, the equation $z = f(x, y)$ defines a surface in the three-dimensional x-y-z space since that equation determines the value of z for any values assigned independently to x and y.

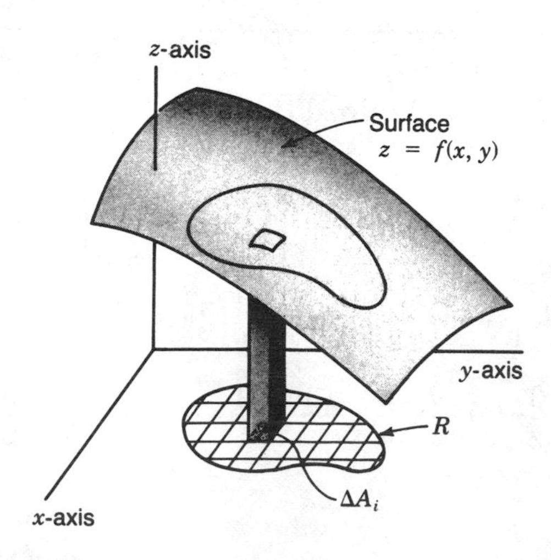

(continued)

We can easily see from the above definition of the double integral that $\iint_R f(x, y)\, dA$ is equal to the volume V of space under the surface $z = f(x, y)$ and above the region R. In this case $f(x_i, y_i)$ is the height of column above ΔA_i. Therefore, $f(x_i, y_i)\, \Delta A$ is approximately equal to the volume of that column. The sum of all these columns is then approximately equal to the volume under the surface. In the limit as $\Delta A_i \to 0$, the sum defining the double integral becomes equal to the volume under the surface and above R, so

$$V = \iint_R z\, dA = \iint_R f(x, y)\, dA.$$

Calculate the volume under surface defined by $z = x + y$ and above the rectangle whose sides are determined by the lines $x = 1, x = 4, y = 0$, and $y = 3$.

Go to **381**.

381

The answer is 36. If you obtained this result, go to frame **382**. If not, study the following.

$$V = \iint_R (x + y)\, dA = \int_1^4 \left[\int_0^3 (x + y)\, dy\right] dx$$

$$= \int_1^4 \left(xy + \frac{y^2}{2}\right)\Big|_1^4 dx = \int_1^4 \left(3x + \frac{9}{2}\right) dx$$

$$= \left(\frac{3}{2}x^2 + \frac{9}{2}x\right)\Big|_1^4 = \frac{(3)(16)}{2} + \frac{(9)(4)}{2} - \frac{3}{2} - \frac{9}{2} = 36.$$

Answer: (378) 90

The iterated integral could just as well have been evaluated in the opposite order.

$$\int_0^3 \left[\int_1^4 (x + y)\, dx\right] dy = \int_0^3 \left(\frac{x^2}{2} + yx\right)\Big|_1^4 dy$$

$$= \int_0^3 \left(\frac{16}{2} + 4y - \frac{1}{2} - y\right) dy = \int_0^3 \left(3y + \frac{15}{2}\right) dx$$

$$= \left(\frac{3}{2}y^2 + \frac{15}{2}y\right)\Big|_0^3 = \frac{27}{2} + \frac{45}{2} = 36.$$

Go to **382**.

382

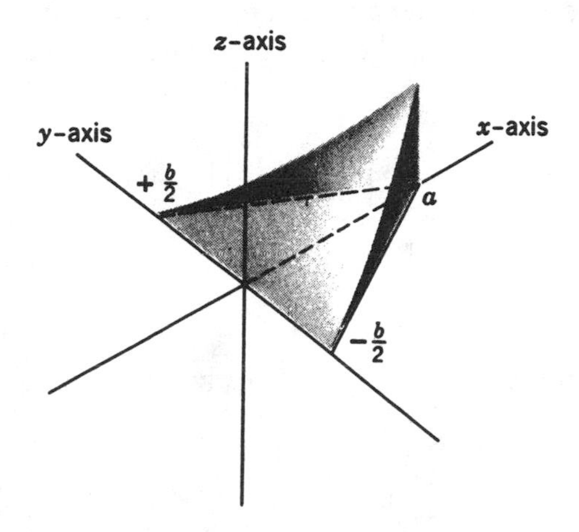

The bottom of this plow-shaped solid is in the form of an isosceles triangle with base b and height a. When oriented along the x-y axes as shown, its thickness is given by $z = Cx^2$, where C is a constant. The problem is to find an expression for the volume.

Volume = ______________.

To check your answer, go to **383**.

383

The volume is $\frac{1}{12}Cba^3$. Read on if you want an explanation; otherwise go to **384**.

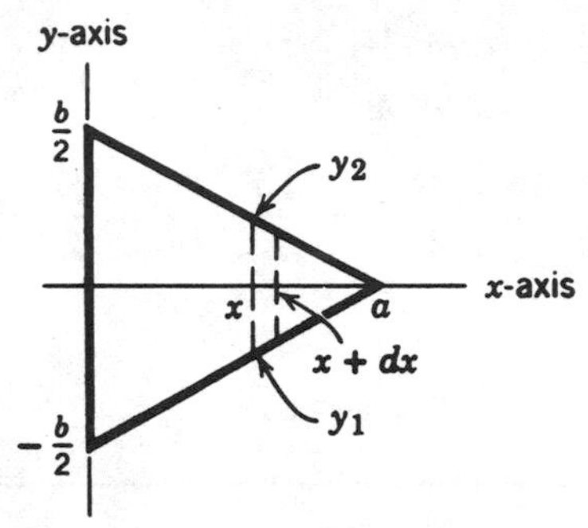

The base of the object forms a triange, as shown. The integral can be carried out with respect to x and y in either order. We shall integrate first over y.

$$V = \int_0^a \left[\int_{y_1}^{y_2} z\, dy\right] dx = \int_0^a \left[\int_{y_1}^{y_2} Cx^2\, dy\right] dx$$

$$= \int_0^a Cx^2 y \Big|_{y_1}^{y_2} dx.$$

From the drawing, $y_2 = \frac{b}{2}\left(1 - \frac{x}{a}\right) = -y_1$, so that $Cx^2 y \Big|_{y_1}^{y_2} = Cx^2 b\left(1 - \frac{x}{a}\right)$, and

$$V = Cb\int_0^2 x^2\left(1 - \frac{x}{a}\right) dx = Cb\left(\frac{1}{3}a^3 - \frac{1}{4}a^3\right) = \frac{1}{12}\, Cba^3.$$

The integral can also be evaluated in reverse order. The calculation is simplified by making use of symmetry; the volume is twice the volume over the upper triangle. Thus

$$V = 2\int_0^{b/2}\left[\int_0^{x(y)} Cx^2 dx\right] dy,$$

where $x(y) = a\left(1 - \frac{2}{b}y\right)$. The answer is the same, $\frac{Cba^3}{12}$.

Go to **384**.

Conclusion

384

Well, here you are at the very last frame. You should get some reward for all your effort—all we can do is promise that there is only one more "go to" left in the book.

At this point you should understand the principles of integration and be able to do some integrals. With practice your repertoire will increase. Don't be afraid to use integral tables—everyone does. You can find quite large tables in

CRC Standard Mathematical Tables, and also *The Handbook of Chemistry and Physics,* CRC Press, Inc., Boca Raton, Florida.
Table of Integrals, Series, and Products, I. S. Gradshteyn and I. W. Ryzhik, Academic Press, New York, 1980.

The next chapter is a review and lists in outline all the ideas presented in the book. Even though you may have already read part of that chapter, you should now study it all. You may also find it is handy for future reference.

The appendixes are crammed full of interesting tidbits: derivations of formulas, explanations of special topics, and the like.

In case you are a glutton for punishment or simply want a little more practice, there is a list of review problems, along with the answers, starting on page 245.

Go to Chapter 4.

CHAPTER FOUR
Review

This chapter is a review and concise summary of what you have learned. Proofs and detailed explanations given in the preceding three chapters are not repeated here; instead, references are given to the appropriate frames. Unlike the rest of the book, this chapter has no questions so it can be read from beginning to end like an ordinary text, except that you may occasionally want to refer back to earlier discussion.

Review of Chapter 1
A FEW PRELIMINARIES

Functions (frames 3–13)

A set is a collection of objects—not necessarily material objects—described in such a way that we have no doubt as to whether a particular object does or does not belong to the set. A set may be described by listing its elements or by a rule.

A *function* is a rule that assigns to each element in a set A one and only one element in a set B. The rule can be specified by a mathematical formula such as $y = x^2$, or by tables of associated numbers. If x is one of the elements of set A, then the element in set B that the function f associates with x is denoted by the symbol $f(x)$, which is usually read as "f of x."

The set A is called the *domain* of the function. The set B of all possible values of $f(x)$ as x varies over the domain is called the *range* of the function.

When a function is defined by a formula such as $f(x) = ax^3 + b$, then x is often called the *independent* variable and $f(x)$ is called the *dependent* variable. Often, however, a single letter is used to represent the single variable as in

$$y = f(x).$$

Here x is the independent variable and y is the dependent variable. In mathematics the symbol x frequently represents an independent variable, f often represents the function, and $y = f(x)$ usually denotes the dependent variable. However any other symbols may be used for the function, the independent variable, and the dependent variable; for example, $x = H(r)$.

Graphs (frames 14–22)

A convenient way to represent a function is to plot a graph as described in frames **15–18**. The mutually perpendicular coordinate axes intersect at the origin. The axis that runs horizontally is called the horizontal axis, or x-axis. The axis that runs vertically is called the vertical axis, or y-axis. Sometimes the value of the x-coordinate of a point is called the abscissa, and the value of the y-coordinate is called the ordinate. In the designation of a typical point by the notation (a, b), we will always designate the x-coordinate first and the y-coordinate second.

The constant function assigns a single fixed number c to each value of the independent variable x. The absolute value function $|x|$ is defined by

$$|x| = \begin{cases} x & \text{if } x \geqslant 0, \\ -x & \text{if } x < 0. \end{cases}$$

Linear and Quadratic Functions (frames 23–39)

An equation of the form $y = mx + b$ where m and b are constants is called *linear* because its graph is a straight line. The slope of a linear function is defined by

$$\text{Slope} = \frac{y_2 - y}{x_2 - x_1} = \frac{y_1 - y_2}{x_1 - x_2}.$$

From the definition it is easy to see (frame **29**) that the slope of the above linear equation is m.

An equation of the form $y = ax^2 + bx + c$, where a, b, and c, are constants, is called a *quadratic equation*. Its graph is called a *parabola*. The values of x at $y = 0$ satisfy $ax^2 + bx + c = 0$ and are called the *roots* of the equation. Not all quadratic equations have real roots. The equation $ax^2 + bx + c = 0$ has two roots given by

$$x = \frac{-b \pm \sqrt{b^2 - 4ac}}{2a}.$$

Trigonometry (frames 40–74)

Angles are measured in either *degrees* or *radians*.

A circle is divided into 360 equal *degrees*. The number of *radians* in an angle is equal to the length of the subtending arc divided by the length of the radius (frame **42**). The relation between degrees and radians is

$$1 \text{ rad} = \frac{360^\circ}{2\pi}.$$

Rotations can be clockwise or counterclockwise. An angle formed by rotating in a counterclockwise direction is taken to be positive.

The trigonometric functions are defined in conjunction with the figure.

The definitions are

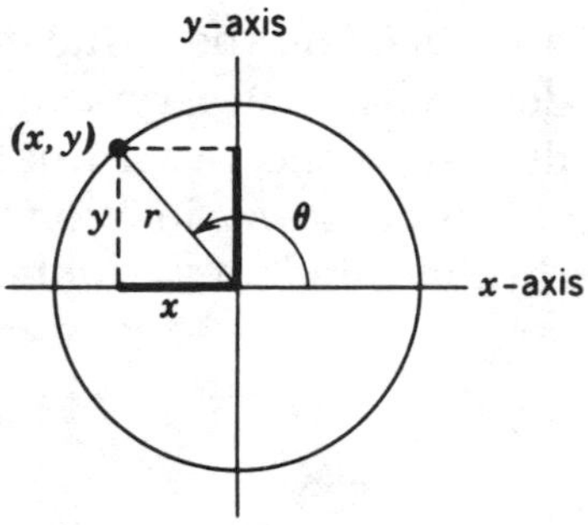

$$\sin\theta = \frac{y}{r}, \qquad \cos\theta = \frac{x}{r},$$

$$\tan\theta = \frac{y}{x}, \qquad \cot\theta = \frac{1}{\tan\theta} = \frac{x}{y},$$

$$\sec\theta = \frac{1}{\cos\theta} = \frac{r}{x}, \qquad \csc\theta = \frac{1}{\sin\theta} = \frac{r}{y}.$$

Although $r = \sqrt{x^2 + y^2}$ is always positive, x and y can be either positive or negative and the above quantities may be positive or negative depending on the value of θ. From the Pythagorean theorem it is easy to see (frame **56**) that

$$\sin^2\theta + \cos^2\theta = 1.$$

The sines and cosines for the sum of two angles are given by:

$$\sin(\theta + \phi) = \sin\theta\cos\phi + \cos\theta\sin\phi,$$
$$\cos(\theta + \phi) = \cos\theta\cos\phi - \sin\theta\sin\phi.$$

The inverse trigonometric function designates the angle for which the trigonometric function has the specified value. Thus the inverse trigonometric function to $y = \sin\theta$ is $\theta = \sin^{-1} y$ and, similar definitions apply to $\cos^{-1} x$, $\tan^{-1} x$, etc. [Warning: This notation is standard, but it can be confusing: $\sin^{-1} x \neq (\sin x)^{-1}$. An older notation for $\sin^{-1} x$ is $\arcsin x$.]

Exponentials and Logarithms (frames 75–95)

If a is multiplied by itself as $aaa \cdots$ with m factors, the product is written as a^m. Furthermore, by definition, $a^{-m} = 1/a^m$. From this it follows that

$$
\begin{aligned}
a^m a^n &= a^{m+n}, \\
\frac{a^m}{a^n} &= a^{m-n}, \\
a^0 &= \frac{a^m}{a^m} = 1, \\
(a^m)^n &= a^{mn}, \\
(ab)^m &= a^m b^m.
\end{aligned}
$$

If $b^n = a$, b is called the nth root of a and is written as $b = a^{1/n}$. If m and n are integers,

$$a^{m/n} = (a^{1/n})^m.$$

The meaning of exponents can be extended to irrational numbers (frame **84**) and the above relations also apply with irrational exponents, so $(a^x)^b = a^{bx}$, etc.

The definition of log x (the logarithm of x to the base 10) is

$$x = 10^{\log x}.$$

The following important relations can easily be seen to apply to logarithms (frame **91**):

$$
\begin{aligned}
\log ab &= \log a + \log b, \\
\log (a/b) &= \log a - \log b, \\
\log a^n &= n \log a.
\end{aligned}
$$

The logarithm of x to another base r is written as $\log_r x$ and is defined by

$$x = r^{\log_r x}.$$

The above three relations for logarithms of a and b are correct for logarithms to any base provided the same base is used for all the logarithms in each equation.

A particular important base is $r = e = 2.71828\ldots$ as defined in frame **109**. Logarithms to the base e are so important in calculus that they are given a different name; they are called *natural* logarithms and written as ln. With this notation the natural logarithm of x is defined by

$$e^{\ln x} = x.$$

If we take the logarithm to base 10 of both sides of the equation,

$$\log e^{\ln x} = \log x,$$

$$\ln x \log e = \log x,$$

$$\ln x = \frac{\log x}{\log e}.$$

Since the numerical value of $1/\log e$ is 2.303 . . .,

$$\ln x = (2.303\ldots)\log x$$

Review of Chapter 2
DIFFERENTIAL CALCULUS

Limits (frames 97–115)

Definition of a Limit: Let $f(x)$ be defined for all x in an interval about $x = a$, but not necessarily at $x = a$. If there is a number L such that to each positive number ϵ there corresponds a positive number δ such that

$$|f(x) - L| < \epsilon \qquad \text{provided } 0 < |x - a| < \delta,$$

we say that L is the *limit* of $f(x)$ as x approaches a, and write

$$\lim_{x\to a} f(x) = L.$$

The ordinary algebraic manipulations can be performed with limits as shown in Appendix A2; thus

$$\lim_{x\to a} [F(x) + G(x)] = \lim_{x\to a} F(x) + \lim_{x\to a} G(x).$$

Two trigonometric limits are of particular interest (Appendix A3):

$$\lim_{\theta\to 0} \frac{\sin \theta}{\theta} = 1 \qquad \text{and} \qquad \lim_{\theta\to 0} \frac{1 - \cos \theta}{\theta} = 0.$$

The following limit is of such great interest in calculus that it is given the special name e, as discussed in frame **109** and Appendix A8:

$$e = \lim_{x\to 0}(1 + x)^{1/x} = 2.71828\ldots$$

Velocity (frames 116–145)

If the function S represents the distance from a fixed location of a point moving at a varying speed along a straight line, the *average velocity* $\bar{v}$ between times t_1 and t_2 is given by

$$\bar{v} = \frac{S_2 - S_1}{t_2 - t_1},$$

whereas the *instantaneous velocity* v (frame **133**) at time t_1 is

$$v = \lim_{t_2\to t_1}\frac{S_2 - S_1}{t_2 - t_1}.$$

This equals the slope at time t_1 of the curve of S plotted in terms of time (frame **131**). It is often convenient to write $S_2 - S_1 = \Delta S$ and $t_2 - t_1 = \Delta t$, so

$$v = \lim_{\Delta t\to 0}\frac{\Delta S}{\Delta t}.$$

Derivatives (frames 146–159)

If $y = f(x)$, the rate of change of y with respect to x is $\lim_{\Delta x\to 0}\frac{\Delta y}{\Delta x}$. The $\lim_{\Delta x\to 0}\frac{\Delta y}{\Delta x}$ is called the *derivative* of y with respect to x and is often written as $\frac{dy}{dx}$ (sometimes it is written y'). Thus

$$y' = \frac{dy}{dx} = \lim_{\Delta x\to 0}\frac{\Delta y}{\Delta x} = \lim_{x_2\to x_1}\frac{y_2 - y_1}{x_2 - x_1} = \lim_{x_2\to x_1}\frac{f(x_2) - f(x_1)}{x_2 - x_1}$$

is the derivative of y with respect to x. The derivative $\frac{dy}{dx}$ is equal to the slope of the curve of y plotted against x.

Graphs of Functions and Their Derivatives (frames 160–169)

From a graph of a function we can obtain the slope of the curve at different points, and by plotting a new curve of the slopes we can determine the general character and qualitative behavior of the derivative. See frames **160–169** for examples.

Differentiation (frames 170–241)

From the definition of the derivative, a number of formulas for differentiation can be derived. We will review just one example here: the method is typical. Let u and v be variables that depend on x.

$$\frac{d(uv)}{dx} = \lim_{\Delta x \to 0} \frac{\Delta(uv)}{\Delta x} = \lim_{\Delta x \to 0} \frac{(u + \Delta u)(v + \Delta v) - uv}{\Delta x}$$

$$\frac{d(uv)}{dx} = \lim_{\Delta x \to 0} \frac{uv + u\,\Delta v + v\,\Delta u + \Delta u\,\Delta v - uv}{\Delta x}$$

$$= u \lim_{\Delta x \to 0} \frac{\Delta v}{\Delta x} + v \lim_{\Delta x \to 0} \frac{\Delta u}{\Delta x} + \lim_{\Delta x \to 0} \frac{\Delta u\,\Delta v}{\Delta x}$$

$$= u\frac{dv}{dx} + v\frac{du}{dx} + 0.$$

The important relations which you should remember are listed here. There is a more complete list in Table 1, page 254. In the following expressions u and v are variables that depend on x, w depends on u, which in turn depends on x, and a and n are constants. All angles are measured in radians.

	Frame
$\frac{da}{dx} = 0$	**172**
$\frac{d}{dx}(ax) = a$	**174**
$\frac{dx^n}{dx} = nx^{n-1}$	**180**
$\frac{d}{dx}(u + v) = \frac{du}{dx} + \frac{dv}{dx}$	**186**
$\frac{d}{dx}(uv) = u\frac{dv}{dx} + v\frac{du}{dx}$	**189**
$\frac{d}{dx}\left(\frac{u}{v}\right) = \frac{1}{v^2}\left(v\frac{du}{dx} - u\frac{dv}{dx}\right)$	**194**
$\frac{dw}{dx} = \frac{dw}{du}\frac{du}{dx}$	**198**

$$\frac{d \sin x}{dx} = \cos x \qquad \textbf{211}$$

$$\frac{d \cos x}{dx} = -\sin x \qquad \textbf{212}$$

$$\frac{d \ln x}{dx} = \frac{1}{x} \qquad \textbf{226}$$

$$\frac{de^x}{dx} = e^x \qquad \textbf{235}$$

In the above list $e = 2.71828\ldots$ and $\ln x$ is the natural logarithm of x defined by $\ln x = \log_e x$.

More complicated functions can ordinarily be differentiated by applying several of the rules in Table 1 successively. Thus

$$\frac{d}{dx}(x^3 + 3x^2 \sin 2x) = \frac{dx^3}{dx} + 3\frac{dx^2}{dx}\sin 2x + 3x^2 \frac{d \sin 2x}{dx}$$

$$= 3x^2 + 6x \sin 2x + 3x^2 \frac{d \sin 2x}{d\,2x}\frac{d\,2x}{dx}$$

$$= 3x^2 + 6x \sin 2x + 6x^2 \cos 2x.$$

Higher-Order Derivatives (frames 242–249)

If we differentiate $\frac{dy}{dx}$ with respect to x, the result is called the *second derivative* of y with respect to x and is written $\frac{d^2y}{dx^2}$. Likewise the nth derivative of y with respect to x is the result of differentiating y n times successively with respect to x and is written $\frac{d^ny}{dx^n}$.

Maxima and Minima (frames 250–261)

If $f(x)$ has a maximum or minimum value for some value of x in an interval, then its derivative $\frac{df}{dx}$ is zero for that x. If in addition $\frac{d^2f}{dx^2} < 0$, $f(x)$ has maximum value. If on the other hand $\frac{d^2f}{dx^2} > 0$, $f(x)$ has a minimum value there.

Differentials (frames 262–272)

If x is an independent variable and $y = f(x)$, the *differential* dx of x is defined as equal to any increment, $x_2 - x_1$, where x_1 is the point of interest. The differential dx can be positive or negative, large or small, as we please. Then dx, like x, is an independent variable. The differential dy is then *defined* by the following rule:

$$dy = y' \, dx.$$

where y' is the derivative of y with respect to x. Although the meaning of the derivative, y', is $\lim_{\Delta x \to 0} \frac{\Delta y}{\Delta x}$, we see that it can now be interpreted as the ratio of the differentials dy and dx. As discussed in frames **265** and **266**, dy is not the same as Δy, though

$$\lim_{dx = \Delta x \to 0} \frac{dy}{\Delta y} = 1.$$

The differentiation formulas can easily be written in terms of differentials. Thus if $y = x^n$,

$$dy = d(x^n) = \frac{d}{dx}(x^n)\, dx = nx^{n-1}\, dx.$$

A useful relation which is implied by the differential notation and discussed further in Appendix A10 is

$$\frac{dx}{dy} = \frac{1}{dy/dx}.$$

Review of Chapter 3
INTEGRAL CALCULUS

The Area under a Curve (frames 290–299)

If $A(x)$ is the area under the curve defined by $y = f(x)$, then

$$\frac{dA(x)}{dx} = A'(x) = f(x).$$

To find $A(x)$ for a given $f(x)$, we then need to find a function whose derivative is equal to $f(x)$. The process of finding a function whose derivative is another function is called *integration* or *antidiferrentiation* and is the subject of the following sections.

Integration (frames 300–308)

The antiderivative $F(x)$ of the function of x is most frequently written in the form

$$F(x) = \int f(x)\, dx.$$

$F(x)$ is usually called the *indefinite integral* of $f(x)$. Since the derivative of a constant is zero, any arbitrary constant c can be added to an indefinite integral and the sum will also be an indefinite integral of the same function $f(x)$. It is important not to omit this constant. Otherwise the answer is incomplete.

Indefinite integrals are often found by hunting for an expression which, when differentiated, gives the integrand $f(x)$. Thus from the earlier result that

$$\frac{d \cos x}{dx} = -\sin x$$

we have that

$$\int \sin x\, dx = -\cos x + c.$$

By starting with known derivatives as in Table 1, a useful list of integrals can be found. Such a list is given in frame **306** and for convenience is repeated in Table 2. You can reconstruct the most important of these formulas from the differentiation expressions in Table 1. More complicated integrals can often be found in large tables, such as those listed in the references on page 207.

Some Techniques of Integration (frames 309–325)

Often an unfamiliar function can be converted into a familiar function having a known integral by using a technique called *change of variable* which is related to the chain rule of differentiation and uses the relation

$$\int w(u)\, dx = \int \left[w(u)\, \frac{du}{dx}\right] dx.$$

Another valuable technique is integration by parts, as described by the relation proved in frame **318**.

$$\int u\, dv = uv - \int v\, du.$$

Frequently a number of different integration procedures are used in a single problem as illustrated in frames **323–325**.

More on the Area under a Curve (frames 326–333)

The area $A(x)$ under a curve defined by the relation $y = f(x)$ can be written as

$$A(x) = F(x) + C,$$

where $F(x)$ is any particular antiderivative of $f(x)$ and C is an arbitrary constant. If we want to know the area bounded by $x = a$ and some value x, the constant c can be evaluated by noting that the area is zero if $x = a$, so

$$A(a) = F(a) + C = 0$$

and $C = -F(a)$. Therefore,

$$A(x) = F(x) - F(a).$$

The area under the curve between $x = a$ and $x = b$ is then

$$A(b) = F(b) - F(a) = F(x)\Big|_a^b$$

where the symbol $F(x)\Big|_a^b$ by definition equals $F(b) - F(a)$.

Definite Integrals (frames 334–352)

An alternative expression for the area A under a curve $f(x)$ between $x = a$ and $x = b$ can be found by dividing the area into N narrow strips parallel to the y-axis, each of area $f(x_i)\,\Delta x$, and summing the strips. In the limit as the width of each strip approaches zero, the limit of the sum approaches the area under the curve. Thus (frame **339**),

$$A = \lim_{\Delta x \to 0} \sum_{i=1}^{N} f(x_i)\,\Delta x.$$

Such a limit is so important that it is given a special name and symbol. It is called the *definite integral* and is written $\int_a^b f(x)\,dx$. Hence by definition.

$$\int_a^b f(x)\,dx = \lim_{\Delta x \to 0} \Sigma f(x_i)\,\Delta x.$$

As a result of this discussion, we see that

$$A = \int_a^b f(x)\,dx.$$

However, we have seen that the area can also be evaluated in terms of the *indefinite integral,*

$$F(x) = \int f(x)\,dx,$$

by

$$A = F(b) - F(a) = F(x)\Big|_a^b = \int f(x)\,dx\,\Big|_a^b.$$

Therefore, by equating the two expressions for A, we have the general evaluation of the *definite* integral in terms of the *indefinite* integral.

$$\int_a^b f(x)\,dx = F(x)\Big|_a^b = \int f(x)\,dx\,\Big|_a^b.$$

This result is often called the fundamental theorem of integral calculus.

Numerical Integration (frames 353–358)

When it is not possible to find an analytic expresion for an integral, the indefinite integral is often evaluated by methods of numerical integration. A particularly effective method is Simpson's rule in which the entire interval over which the integral is to be evaluated is divided into an even number N of equal intervals of width Δ. By Simpson's rule the numerical value of the integral is approximately given by

$$\int_A^B y\,dx = \frac{\Delta}{3}(y_0 + 4y_1 + 2y_2 + 4y_3 + 2y_4 + \cdots + 2y_{N-2} + 4y_{N-1} + y_N)$$

The accuracy of the approximation can be increased by increasing N with a corresponding decrease in Δ, but with a corresponding increase in numerical work.

Some Applications of Integration (frames 359–374)

If we know $v(t)$, the *velocity* of a particle as a function of t, we can obtain the *position* of the particle as a function of time by integration. We saw earlier that

$$v = \frac{dS}{dt}$$

so

$$dS = v\,dt$$

and if we integrate both sides of the equation from the initial point ($t = t_0$, $S = 0$) to the final point (t, S), we have

$$S = \int_{t_0}^{t} v\,dt.$$

Applications of integration in finding volumes of symmetric solids are given in frames **366–374**.

Multiple Integrals (frames 375–383)

Multiple integrals may be defined for an arbitrary number of independent variables. We discuss two variables since the procedures for an arbitrary number are merely generalizations of those that apply to two independent variables. The double integral over a region R in the x-y plane of the function $f(x, y)$ is defined as

$$\iint_R f(x, y)\,dA = \lim_{\Delta A_i \to 0} \sum_{i=1}^{N} f(x_i, y_i)\,\Delta A_i,$$

as discussed in frame **376**. The double integral can often be evaluated in terms of the *iterated integral:*

$$\iint_R f(x, y)\,dA = \int_a^b \left[\int_{y_1(x)}^{y_2(x)} f(x, y)\,dy\right] dx.$$

Conclusion (frame 384)

You are now finished. Congratulations! You don't need to do any more work to complete this book. However, if you skipped some of the proofs in Appendix A, we suggest you read them now. You may also want to study some of the additional topics that are described in Appendix B. Finally, if you would like to have some more practice, you should try some of the review problems starting on page 245.

Good luck!

APPENDIX A
Derivations

In this appendix derivations are given of certain of the formulas and theorems not derived earlier.

Appendix A1
TRIGONOMETRIC FUNCTIONS OF SUMS OF ANGLES

A formula can easily be derived for the sine of the sum of two angles, θ and ϕ, with the aid of the drawing in which the radius of the circle is unity.

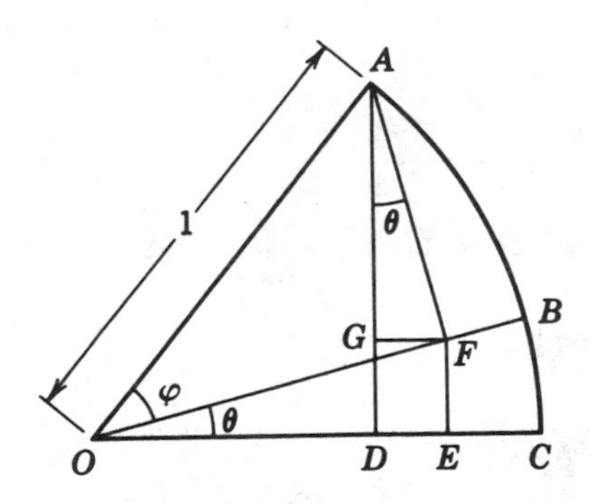

$$\begin{aligned}\sin(\theta + \phi) &= AD = FE + AG \\ &= OF \sin\theta + AF \cos\theta \\ &= \sin\theta \cos\phi + \cos\theta \sin\phi.\end{aligned}$$

In a similar fashion with the same figure,

$$\begin{aligned}\cos(\theta + \phi) &= OD = OE - DE \\ &= OF \cos\theta - AF \sin\theta \\ &= \cos\theta \cos\phi - \sin\theta \sin\phi.\end{aligned}$$

Appendix A2
SOME THEOREMS ON LIMITS

In this appendix we shall prove several useful theorems on limits. These theorems will show that the usual algebraic manipulations can be done with expressions involving limits. We shall show, for example, that

$$\lim_{x \to a} [F(x) + G(x)] = \lim_{x \to a} F(x) + \lim_{x \to a} G(x).$$

Although such results are intuitively reasonable, they require a formal proof.

Before deriving these theorems, we need to note some general properties of the absolute value function introduced in frame **20**. These properties are

$$|a + b| \leqslant |a| + |b|, \tag{1}$$
$$|ab| = |a| \times |b|. \tag{2}$$

It is easy to see that these relations are true by considering in turn each of all the possible cases: a and b both negative, both positive, of opposite sign, and one or both equal to zero.

We are now ready to discuss theorems on limits which apply to any two functions F and G such that

$$\lim_{x \to a} F(x) = L \qquad \text{and} \qquad \lim_{x \to a} G(x) = M.$$

Theorem 1

$$\lim_{x \to a} [F(x) + G(x)] = \lim_{x \to a} F(x) + \lim_{x \to a} G(x).$$

Proof: By Equation (1),

$$\begin{aligned} |F(x) + G(x) - (L + M)| &= |[F(x) - L] + [G(x) - M]| \\ &\leqslant |F(x) - L| + |G(x) - M|. \end{aligned}$$

Using the definition of the limit (frame **105**) we see that for any positive number ϵ we can find a positive number δ such that

$$|F(x) - L| < \frac{\epsilon}{2} \qquad \text{and} \qquad |G(x) - M| < \frac{\epsilon}{2}$$

provided $0 < |x - a| < \delta$. (At first sight this may appear to differ from the definition of the limit since the symbol ϵ instead of $\epsilon/2$ appeared there. However, the statements apply for any positive number and $\epsilon/2$ is also a positive number.)

The above equations may be combined to give

$$| F(x) + G(x) - (L + M) | < \frac{\epsilon}{2} + \frac{\epsilon}{2} = \epsilon.$$

Therefore, by the definition of the limit in frame **105**,

$$\lim_{x \to a} [F(x) + G(x)] = L + M = \lim_{x \to a} F(x) + \lim_{x \to a} G(x).$$

Theorem 2

$$\lim_{x \to a} [F(x)\, G(x)] = [\lim_{x \to a} F(x)]\, [\lim_{x \to a} G(x)].$$

Proof: The proof is somewhat similar to the preceding. By writing out all the terms, we can see that the following is true identically:

$$F(x)\, G(x) - LM = [F(x) - L]\, [G(x) - M] + L[G(x) - M] + M[F(x) - L].$$

Therefore, by Equation (1),

$$| F(x)\, G(x) - LM | \\ \leqslant | [F(x) - L]\, [G(x) - M] | + | L[G(x) - M] | + | M[F(x) - L] |.$$

Let ϵ be any positive number less than 1. Then by the meaning of limits we can find a positive number δ such that if $0 < | x - a | < \delta$,

$$| F(x) - L | < \frac{\epsilon}{2}, \qquad | L[G(x) - M] | < \frac{\epsilon}{4}, \qquad | M[F(x) - L] | < \frac{\epsilon}{4},$$

and

$$| G(x) - M | < \frac{\epsilon}{2}.$$

Then

$$| F(x)\, G(x) - LM | < \frac{\epsilon^2}{4} + \frac{\epsilon}{4} + \frac{\epsilon}{4} = \frac{\epsilon^2}{4} + \frac{\epsilon}{2} \leqslant \frac{\epsilon}{4} + \frac{\epsilon}{2} = \frac{3}{4}\,\epsilon,$$

where the next to the last step arises as a result of our earlier restriction to $\epsilon < 1$.

Consequently,

$$| F(x)\, G(x) - LM | < \epsilon$$

so by the definition of the limit,

$$\lim_{x \to a} [F(x)\, G(x)] = LM = [\lim_{x \to a} F(x)]\, [\lim_{x \to a} G(x)].$$

Theorem 3

$$\lim_{x\to a}\frac{F(x)}{G(x)}=\frac{\lim\limits_{x\to a}F(x)}{\lim\limits_{x\to a}G(x)} \qquad \text{provided } \lim_{x\to a}G(x)\neq 0.$$

Proof: Since $\lim\limits_{x\to a}G(x)\neq 0$, we can select a value of δ sufficiently small that $G(x)\neq 0$ for $0<|x-a|<\delta$. Then we can write

$$\lim_{x\to a}F(x)=\lim_{x\to a}\left[G(x)\frac{F(x)}{G(x)}\right]=\lim_{x\to a}G(x)\lim_{x\to a}\frac{F(x)}{G(x)}$$

$$=M\lim_{x\to a}\frac{F(x)}{G(x)}$$

where $M=\lim\limits_{x\to a}G(x)$.

Therefore, since $M\neq 0$, we have

$$\lim_{x\to a}\frac{F(x)}{G(x)}=\frac{\lim\limits_{x\to a}F(x)}{M}=\frac{\lim\limits_{x\to a}F(x)}{\lim\limits_{x\to a}G(x)}.$$

Note that if $M=0$ this expression is meaningless and we must evaluate $\frac{F(x)}{G(x)}$ before taking the limit.

Appendix A3
LIMITS INVOLVING TRIGONOMETRIC FUNCTIONS

1. Proof that

$$\lim_{\theta\to 0}\frac{\sin\theta}{\theta}=1.$$

To prove this, draw an arc of a unit circle as shown such that $AB=AE=1$ and $\theta=\angle EAB$. Geometrically it is apparent that area $ADE\geqslant$ area $ABE\geqslant ABC$. Therefore ½ $(\overline{AE})(\overline{DE})\geqslant$ area $ABE\geqslant$ ½ $(\overline{AC})(\overline{BC})$. (The symbol $\overline{AE}$ represents the length of the straight line segment between A and E.)

Since the area of the circle is π, we have

$$\text{Area } ABE=\pi\frac{\theta}{2\pi}=\frac{1}{2}\theta.$$

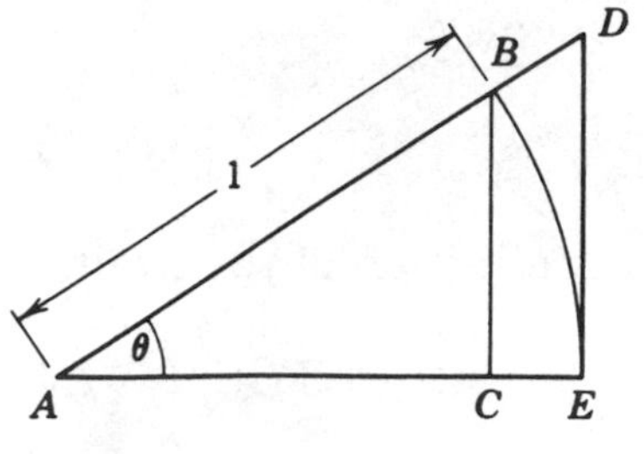

Using the fact that $\overline{DE} = \tan \theta$, we obtain

$$\tfrac{1}{2} \tan \theta \geqslant \tfrac{1}{2} \theta \geqslant \tfrac{1}{2} \cos \theta \sin \theta.$$

Dividing through by $\tfrac{1}{2} \sin \theta$ yields

$$\frac{1}{\cos \theta} \geqslant \frac{\theta}{\sin \theta} \geqslant \cos \theta.$$

Take the reciprocals of this expression. Since the reciprocal of a large number is smaller than the reciprocal of a small number (providing both numbers are positive), this operation reverses the order of the inequality:

$$\cos \theta \leqslant \frac{\sin \theta}{\theta} \leqslant \frac{1}{\cos \theta}.$$

So

$$\lim_{\theta \to 0} \cos \theta \leqslant \lim_{\theta \to 0} \frac{\sin \theta}{\theta} \leqslant \lim_{\theta \to 0} \frac{1}{\cos \theta}$$

and

$$1 \leqslant \lim_{\theta \to 0} \frac{\sin \theta}{\theta} \leqslant 1.$$

Therefore,

$$\lim_{\theta \to 0} \frac{\sin \theta}{\theta} = 1.$$

2. Proof that

$$\lim_{\theta \to 0} \frac{1 - \cos \theta}{\theta} = 0.$$

This can be proved as follows:

$$1 - \cos \theta = \frac{(1 - \cos \theta)(1 + \cos \theta)}{1 + \cos \theta} = \frac{1 - \cos^2 \theta}{1 + \cos \theta}$$

$$= \frac{\sin^2 \theta}{1 + \cos \theta} \leqslant \sin^2 \theta \qquad \text{for } 0 \leqslant \theta < \frac{\pi}{2}.$$

Therefore in this limit

$$\frac{1-\cos\theta}{\theta} \leqslant \frac{\sin^2\theta}{\theta}.$$

We then have

$$\lim_{\theta\to 0}\frac{1-\cos\theta}{\theta} \leqslant \left(\lim_{\theta\to 0}\frac{\sin\theta}{\theta}\right)\left(\lim_{\theta\to 0}\sin\theta\right) = 1\times 0 = 0.$$

But for all positive θ, $0 \leqslant \frac{1-\cos\theta}{\theta}$. Hence,

$$0 \leqslant \lim_{\theta\to 0}\frac{1-\cos\theta}{\theta} \leqslant 0.$$

The only way to satisfy both of these conditions is for

$$\lim_{\theta\to 0}\frac{1-\cos\theta}{\theta} = 0.$$

Appendix A4
DIFFERENTIATION OF x^n

Consider first the case of n a positive integer.

$$\begin{aligned} y &= x^n, \\ y + \Delta y &= (x + \Delta x)^n. \end{aligned} \tag{1}$$

The right side can be expanded by the binomial theorem (if you are not familiar with this, look it up in any good algebra text) to give

$$\begin{aligned} y + \Delta y &= (x + \Delta x)^n \\ &= x^n + nx^{n-1}\,\Delta x + \frac{n(n-1)}{1\cdot 2}x^{n-2}\,\Delta x^2 + \cdots + \Delta x^n. \end{aligned} \tag{2}$$

If we subtract Equation (1) from Equation (2) and divide by Δx, we have

$$\frac{\Delta y}{\Delta x} = nx^{n-1} + \frac{n(n-1)}{1\cdot 2}x^{n-2}\,\Delta x + \cdots + \Delta x^{n-1}.$$

Therefore,

$$\frac{dy}{dx} = \lim_{\Delta x \to 0} \frac{\Delta y}{\Delta x} = nx^{n-1}.$$

Although the above theorem has been proved only for n being a positive integer, we can also show it is true for $n = 1/q$ where q is a positive integer. Let

$$y = x^{1/q}$$

so

$$x = y^q.$$

By the preceding theorem, then,

$$\frac{dx}{dy} = qy^{q-1}.$$

But by Appendix A11,

$$\frac{dy}{dx} = \frac{1}{dx/dy} = \frac{1}{qy^{q-1}} = \frac{1}{q}y^{1-q} = \frac{1}{q}(x^{1/q})^{1-q}$$

$$\frac{dy}{dx} = \frac{1}{q}\,x^{(1/q)-1} = nx^{n-1}.$$

We can further see that this theorem holds for $n = p/q$ where p and q are both positive integers.

$$y = x^n = x^{p/q}.$$

Let

$$w = x^{1/q}$$

so

$$y = w^p.$$

Then

$$\frac{dy}{dx} = \frac{dy}{dw}\frac{dw}{dx} = pw^{p-1}\left(\frac{1}{q}\right)x^{(1/q)-1} = px^{(p/q)-(1/q)}\left(\frac{1}{q}\right)x^{(1/q-1)}$$

$$= \left(\frac{p}{q}\right)x^{(p/q)-1} = nx^{n-1}.$$

So far we have seen that the rule for differentiating x^n applies if n is any positive fraction. We will now see that it applies for negative fractions as well. Let $n = -m$, where m is a positive fraction. Then

$$\frac{d(x^n)}{dx} = \frac{d(x^{-m})}{dx} = \frac{d}{dx}\left(\frac{1}{x^m}\right) = -\frac{dx^m/dx}{(x^m)^2}$$

$$= -\frac{mx^{m-1}}{x^{2m}} = (-\mathrm{m})x^{-m-1} = nx^{n-1}.$$

Up to now our discussion applies if n is any rational number. However, the result may be extended to any irrational real number by the methods used in frame **84**. Therefore, for any real number n, whether rational or irrational, and regardless of sign,

$$\frac{d(x^n)}{dx} = nx^{n-1}.$$

Appendix A5
DIFFERENTIATION OF TRIGONOMETRIC FUNCTIONS

From Appendix A1,

$$\frac{d \sin \theta}{d\theta} = \lim_{\Delta\theta \to 0} \frac{\sin(\theta + \Delta\theta) - \sin \theta}{\Delta\theta}$$

$$= \lim_{\Delta\theta \to 0} \frac{\sin \theta \cos \Delta\theta + \cos \theta \sin \Delta\theta - \sin \theta}{\Delta\theta}$$

$$= \sin \theta \lim_{\Delta\theta \to 0} \frac{\cos \Delta\theta - 1}{\Delta\theta} + \cos \theta \lim_{\Delta\theta \to 0} \frac{\sin \Delta\theta}{\Delta\theta}.$$

The two limits were evaluated in Appendix A3 as 0 and 1, respectively, so

$$\frac{d \sin \theta}{d\theta} = \cos \theta.$$

Likewise,

$$\begin{aligned}\frac{d\cos\theta}{d\theta} &= \lim_{\Delta\theta\to 0}\frac{\cos(\theta+\Delta\theta)-\cos\theta}{\Delta\theta}\\ &= \lim_{\Delta\theta\to 0}\frac{\cos\theta\cos\Delta\theta-\sin\theta\sin\Delta\theta-\cos\theta}{\Delta\theta}\\ &= \cos\theta\lim_{\Delta\theta\to 0}\frac{\cos\Delta\theta-1}{\Delta\theta}-\sin\theta\lim_{\Delta\theta\to 0}\frac{\sin\Delta\theta}{\Delta\theta}\\ &= -\sin\theta.\end{aligned}$$

Derivatives of other trigonometric functions can be found by expressing them in terms of sines and cosines, as in Chapter 2.

Appendix A6
DIFFERENTIATION OF THE PRODUCT OF TWO FUNCTIONS

Let $y = uv$, where u and v are variables which depend on x. Then

$$y + \Delta y = (u + \Delta u)(v + \Delta v) = uv + u\,\Delta v + v\,\Delta u + \Delta u\,\Delta v.$$

Then

$$\begin{aligned}\frac{dy}{dx} &= \lim_{\Delta x\to 0}\frac{(y+\Delta y)-y}{\Delta x} = \lim_{\Delta x\to 0}\frac{(uv+u\Delta v+v\Delta u+\Delta u\Delta v)-uv}{\Delta x}\\ &= \lim_{\Delta x\to 0}\left(u\frac{\Delta v}{\Delta x}+v\frac{\Delta u}{\Delta x}+\Delta u\frac{\Delta v}{\Delta x}\right).\end{aligned}$$

But

$$\lim_{\Delta x\to 0}\Delta u\frac{\Delta v}{\Delta x} = \left(\lim_{\Delta x\to 0}\Delta u\right)\times\left(\lim_{\Delta x\to 0}\frac{\Delta v}{\Delta x}\right) = 0\times\frac{dv}{dx} = 0,$$

where we have used Theorem 2 of Appendix A2. Thus

$$\frac{dy}{dx} = u\lim_{\Delta x\to 0}\frac{\Delta v}{\Delta x}+v\lim_{\Delta x\to 0}\frac{\Delta u}{\Delta x} = u\frac{dv}{dx}+v\frac{du}{dx}.$$

Appendix A7
CHAIN RULE FOR DIFFERENTIATION

Let $w(u)$ depend on u, which in turn depends on x. Then

$$\Delta w = w(u + \Delta u) - w(u)$$

so

$$\frac{\Delta w}{\Delta x} = \frac{\Delta w}{\Delta u}\frac{\Delta u}{\Delta x} = \frac{w(u + \Delta u) - w(u)}{\Delta u}\frac{\Delta u}{\Delta x}.$$

Therefore, using Theorem 2 of Appendix A1, we have

$$\frac{dw}{dx} = \lim_{\Delta x \to 0}\frac{\Delta w}{\Delta x} = \lim_{\Delta x \to 0}\frac{\Delta w}{\Delta u}\lim_{\Delta x \to 0}\frac{\Delta u}{\Delta x} = \left(\frac{dw}{du}\right)\left(\frac{du}{dx}\right).$$

Appendix A8
DIFFERENTIATION OF ln x

Let

$$y = \ln x,$$
$$y + \Delta y = \ln(x + \Delta x).$$

Then

$$\frac{\Delta y}{\Delta x} = \frac{y + \Delta y - y}{\Delta x} = \frac{\ln(x + \Delta x) - \ln x}{\Delta x}.$$

From frame **91**,

$$\frac{\Delta y}{\Delta x} = \frac{1}{\Delta x}\ln\left(\frac{x + \Delta x}{x}\right) = \frac{1}{x}\frac{x}{\Delta x}\ln\left(1 + \frac{\Delta x}{x}\right)$$
$$= \frac{1}{x}\ln\left(1 + \frac{\Delta x}{x}\right)^{x/\Delta x} = \frac{1}{x}\ln(1 + l)^{1/l}$$

where we have written l for $\frac{\Delta x}{x}$. Note that as $\Delta x \to 0$, $l \to 0$. Therefore,

$$\frac{dy}{dx} = \lim_{\Delta x \to 0} \frac{\Delta y}{\Delta x} = \lim_{\Delta x \to 0} \left[\frac{1}{x} \ln(1 + l)^{1/l}\right]$$

$$= \frac{1}{x} \ln\left[\lim_{l \to 0}(1 + l)^{1/l}\right]$$

$$= \frac{1}{x} \ln e = \frac{1}{x}$$

since $\ln e = \log_e e = 1$.

Appendix A9
DIFFERENTIALS WHEN BOTH VARIABLES DEPEND ON A THIRD VARIABLE

The relation $dw = \frac{dw}{du}\, du$ is true even when both w and u depend on a third variable. To prove this, let both u and w depend on x. Then

$$dw = \frac{dw}{dx}\, dx \qquad \text{and} \qquad du = \frac{du}{dx}\, dx. \tag{1}$$

By the chain rule for differentiating,

$$\frac{dw}{dx} = \left(\frac{dw}{du}\right)\left(\frac{du}{dx}\right),$$

and multiplying through by dx, we have

$$\frac{dw}{dx}\, dx = \left(\frac{dw}{du}\right)\left(\frac{du}{dx}\right) dx,$$

so by Equation (1),

$$dw = \frac{dw}{du}\, du.$$

This theorem justifies the use of the differential notation since it shows that with the differential notation the *chain rule* takes the form of an algebraic identity

$$\frac{dw}{dx} = \frac{dw}{du}\frac{du}{dx}.$$

Appendix A10
PROOF THAT $\dfrac{dy}{dx} = \dfrac{1}{dx/dy}$

If a function is specified by an equation $y = f(x)$, it is ordinarily possible, for at least limited intervals of x, to reverse the roles of the dependent and independent variables and to allow the equation to determine the value of x for a given value of y. (This cannot always be done as in the case of the equation $y = a$, where a is a constant.) When such an inversion is possible, the two derivatives are related by

$$\frac{dy}{dx} = \frac{1}{dx/dy}.$$

This relation can be seen as follows:

$$\frac{dy}{dx} = \lim_{\Delta x \to 0} \frac{\Delta y}{\Delta x} = \lim_{\Delta x \to 0} \frac{1}{\Delta x/\Delta y} = \frac{1}{\lim\limits_{\Delta x \to 0} (\Delta x/\Delta y)}$$

by the limit theorems of Appendix 5. Furthermore, if $\lim\limits_{\Delta x \to 0} \dfrac{\Delta y}{\Delta x} \neq 0$, then $\Delta y \to 0$ as $\Delta x \to 0$, so

$$\frac{dy}{dx} = \frac{1}{\lim\limits_{\Delta y \to 0} (\Delta x/\Delta y)} = \frac{1}{dx/dy}.$$

This result is a further justification of the use of differential notation since normal arithmetic manipulation with differential notation immediately gives

$$\frac{dy}{dx} = \frac{1}{dx/dy}.$$

Appendix A11
PROOF THAT IF TWO FUNCTIONS HAVE THE SAME DERIVATIVE THEY DIFFER ONLY BY A CONSTANT

Let the functions be f and g.
Then

$$\frac{d\,f(x)}{dx} = \frac{d\,g(x)}{dx}$$

so

$$\frac{d}{dx}[f(x) - g(x)] = 0.$$

Therefore

$$f(x) - g(x) = C$$

where C is a constant.

This proof depends on the assumption that if $\frac{d\,h(x)}{dx} = 0$, then $h(x)$ is a constant. This is indeed very plausible since the graph of the function $h(x)$ must always have zero slope and hence it should be a straight line parallel to the origin, i.e., $h(x) = C$. A more complicated analytic proof of this theorem is given in advanced books on calculus.

APPENDIX B
Additional Topics

This appendix gives brief discussions of some additional topics in calculus.

Appendix B1
IMPLICIT DIFFERENTIATION

Most of the functions we use in this book can be written in the simple form $y = f(x)$, but this is not always the case. Sometimes we have two variables related by an equation of the form $f(x, y) = 0$. [$f(x, y)$ means that the value of f depends on both x and y.] Here is an example: $x^2y + (y + x)^3 = 0$. We cannot easily solve this equation to yield a result of the form $y = g(x)$, or even $x = h(y)$. However, we can find y' by using the following procedure.

Differentiate both sides of the equation with respect to x, remembering that y depends on x.

$$\frac{d}{dx}(x^2y) + \frac{d}{dx}(y + x)^3 = \frac{d}{dx}(0) = 0,$$

$$x^2\frac{dy}{dx} + 2xy + 3(y + x)^2\left(\frac{dy}{dx} + 1\right) = 0,$$

$$\frac{dy}{dx}[(x^2 + 3(y + x)^2] = -2xy - 3(y + x)^2,$$

$$\frac{dy}{dx} = -\frac{2xy + 3(y + x)^2}{x^2 + 3(y + x)^2}.$$

A function defined by $f(x, y) = 0$ is called an *implicit* function since it implicitly determines the dependence of y on x (or, for that matter, the dependence of x on y in case we need to regard y as the independent variable). The process we have just used, differentiating each term of the equation $f(x, y) = 0$ with respect to the variable of interest, is called *implicit differentiation.*

Here is another example of implicit differentiation. Let $x^2 + y^2 = 1$. The problem is to find y'. We will do this first by implicit differentiation, and then by solving the equation for y and using the normal procedure.

By differentiating both sides of the equation with respect to x, we obtain

$$2x + 2y\ y' = 0.$$

Hence,

$$y' = -\frac{2x}{2y} = -\frac{x}{y}.$$

Alternatively, we can solve for y.

$$y^2 = 1 - x^2, \qquad y = \mp\sqrt{1 - x^2},$$

$$y' = \pm\left(\frac{-2x}{\sqrt{1 - x^2}} \times \frac{1}{2}\right) = \mp\frac{x}{\sqrt{1 - x^2}} = -\frac{x}{y}.$$

We did not need to use implicit differentiation here since we could write the function in the form $y = f(x)$. Often, however, this cannot be done, as in the first example, and implicit differentiation is then necessary.

Appendix B2
DIFFERENTIATION OF THE INVERSE TRIGONOMETRIC FUNCTIONS

1. Evaluation of $\dfrac{d}{dx}(\sin^{-1} x)$.

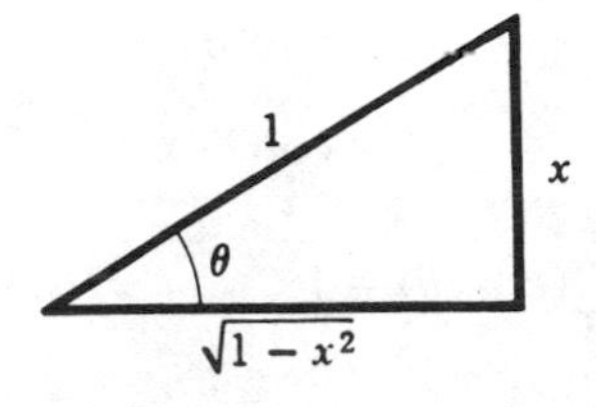

The angle θ is shown inscribed in a right triangle having unit hypotenuse, and an opposite side of length x. Therefore, $\sin\theta = x/1 = x$

and $\theta = \sin^{-1} x$. Differentiating the first expression with respect to x yields.

$$\frac{d \sin \theta}{dx} = 1.$$

Using the chain rule, we have

$$\frac{d}{dx}(\sin \theta) = \frac{d}{d\theta}(\sin \theta)\frac{d\theta}{dx} = \cos \theta \frac{d\theta}{dx} = 1.$$

Then,

$$\frac{d\theta}{dx} = \frac{d}{dx}(\sin^{-1} x) = \frac{1}{\cos \theta}.$$

We can substitute the value $\cos \theta = \sqrt{1 - x^2}$ to obtain our final result:

$$\frac{d}{dx}(\sin^{-1} x) = \frac{1}{\sqrt{1 - x^2}}.$$

Note that we must take the sign of $\sqrt{1 - x^2}$ to agree with that of $\cos \theta$.

2. Evaluation of $\frac{d}{dx}(\cos^{-1} x)$.

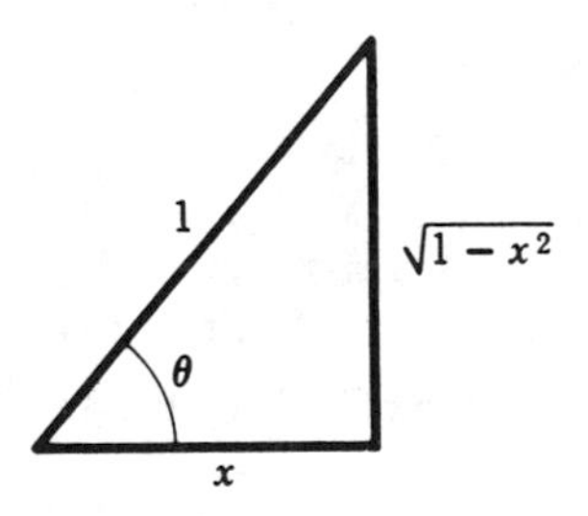

Using the triangle shown and the above procedure, we have

$$x = \cos \theta$$

$$\theta = \cos^{-1} x$$

$$\frac{d}{dx}(\cos \theta) = 1, \qquad \frac{d}{d\theta}(\cos \theta)\frac{d\theta}{dx} = 1,$$

$$\frac{d}{dx}(\cos^{-1} x) = \frac{-1}{\sin \theta} = \frac{-1}{\sqrt{1 - x^2}}.$$

3. Evaluation of $\frac{d}{dx}(\tan^{-1} x)$.

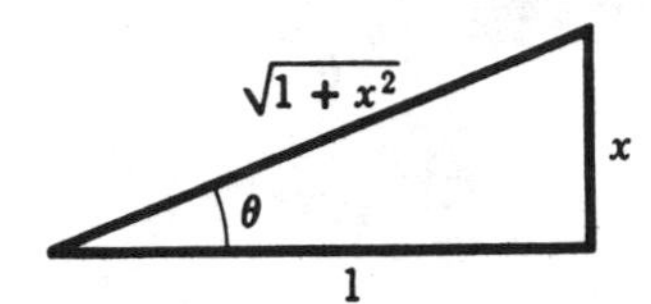

In the triangle shown, $\tan\theta = x$, so that $\theta = \tan^{-1} x$.

$$\frac{d}{dx}(\tan\theta) = \frac{d}{d\theta}(\tan\theta)\frac{d\theta}{dx} = 1.$$

But

$$\frac{d}{d\theta}(\tan\theta) = \sec^2\theta.$$

So

$$\frac{d\theta}{dx} = \frac{1}{\sec^2\theta} = \cos^2\theta = \frac{1}{1+x^2}.$$

$$\frac{d}{dx}(\tan^{-1} x) = \frac{1}{1+x^2}.$$

4. Evaluation of $\frac{d}{dx}(\cot^{-1} x)$.

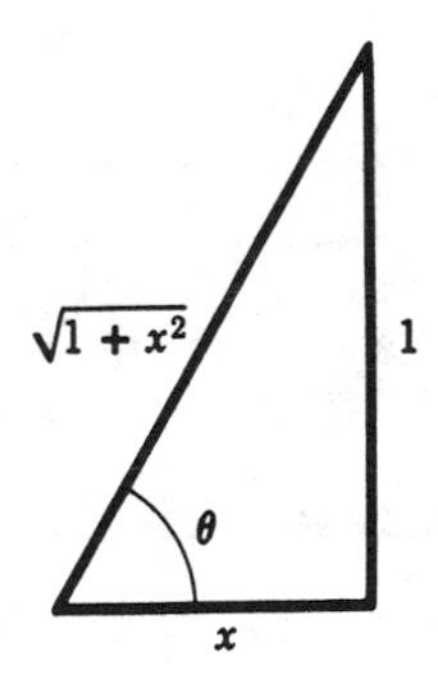

Here $\cot\theta = x$, so that $\theta = \cot^{-1} x$.

$$\frac{d}{dx}(\cot\theta) = \frac{d}{d\theta}(\cot\theta)\frac{d\theta}{dx} = 1.$$

But

$$\frac{d}{d\theta}(\cot\theta) = -\csc^2\theta,$$

so

$$\frac{d\theta}{dx} = -\frac{1}{\csc^2 \theta} = -\sin^2 \theta = \frac{-1}{1 + x^2}.$$

$$\frac{d}{dx}(\cot^{-1} x) = \frac{-1}{1 + x^2}.$$

Appendix B3
PARTIAL DERIVATIVES

In this book we have almost exclusively considered functions defined for a single independent variable. Often, however, two or more independent variables are required to define the function; in this case we have to modify the idea of a derivative. As a simple example, suppose we consider the area of a rectangle A, which is the product of its width w and length l. Thus, $A = f(l, w)$ (read "f of l and w"), where $f(l, w)$ is here $l \times w$. In this discussion we will let l and w vary independently, so they both can be treated as independent variables.

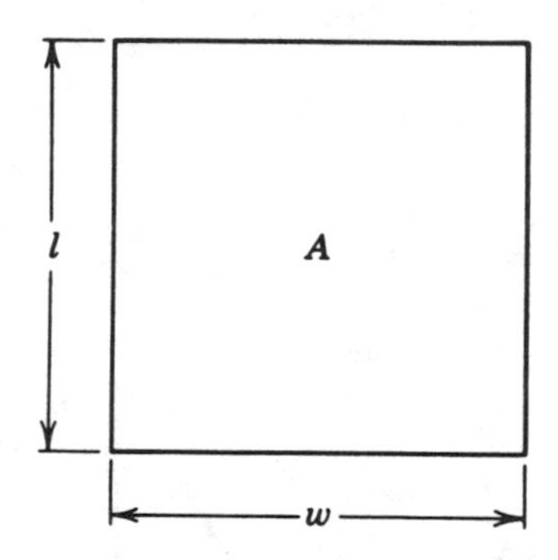

If one of the variables, say w, is temporarily kept constant, then A depends on a single variable, and the rate of change of A with respect to l is simply $\frac{dA}{dl}$. However, because A really depends on two variables, we must modify the definition of the derivative.

$$\text{The rate of change of } A \text{ with respect to } l = \lim_{\Delta l \to 0} \frac{f(l + \Delta l, w) - f(l, w)}{\Delta l}$$

where it is understood that w is held constant as the limit is taken. The above quantity is called the *partial derivative* of A with respect to l and is written $\frac{\partial A}{\partial l}$. In other words the partial derivative is defined by

$$\frac{\partial A}{\partial l} = \frac{\partial f(l, w)}{\partial l} = \lim_{\Delta l \to 0} \frac{f(l + \Delta l, w) - f(l, w)}{\Delta l}.$$

In our example,

$$\frac{\partial A}{\partial l} = \lim_{\Delta l \to 0} \frac{(l + \Delta l) \times w - l \times w}{\Delta l} = w.$$

Similarly,

$$\frac{\partial A}{\partial w} = \lim_{\Delta w \to 0} \frac{f(l, w + \Delta w) - f(l, w)}{\Delta w}$$

$$= \lim_{\Delta w \to 0} \frac{l \times (w + \Delta w) - l \times w}{\Delta w} = l.$$

The differential of A due to changes in l and w of dl and dw, respectively, is by definition

$$dA = \frac{\partial A}{\partial l} dl + \frac{\partial A}{\partial w} dw.$$

By analogy with the argument in 266, it should be plausible that as $dl \to 0$, the increment in A, $\Delta A = f(l + \Delta l, w + \Delta w) - f(l, w)$, approaches dA.

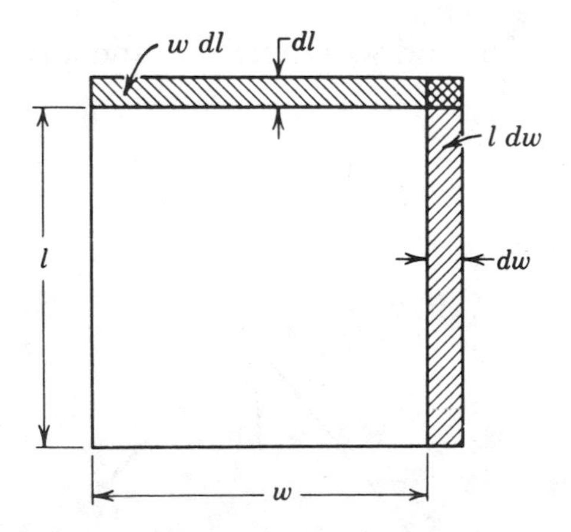

This result is shown by the figure. ΔA is the total increase in area due to dl and dw and comprises all the shaded areas.

$$dA = \frac{\partial A}{\partial l}\, dl + \frac{\partial A}{\partial w}\, dw = w\, dl + l\, dw$$

ΔA and dA differ by the area of the small rectangle in the upper right-hand corner. As $dl \to 0$, $dw \to 0$, the difference becomes negligible compared with the area of each strip.

The above discussion can be generalized to functions depending on any number of variables. For instance, let p depend on $q, r, s, \ldots$.

$$dp = \frac{\partial p}{\partial q}\, dq + \frac{\partial p}{\partial r}\, dr + \frac{\partial p}{\partial s}\, ds + \cdots$$

Here is an example:

$$p = q^2 r \sin z$$

$$\frac{\partial p}{\partial q} = 2qr \sin z$$

$$\frac{\partial p}{\partial r} = q^2 \sin z$$

$$\frac{\partial p}{\partial z} = q^2 r \cos z$$

$$dp = 2qr \sin z\, dq + q^2 \sin z\, dr + q^2 r \cos z\, dz.$$

Here is another example:

The volume of a pyramid with height h and a rectangular base with dimensions l and w is

$$V = \frac{1}{3}\, lwh.$$

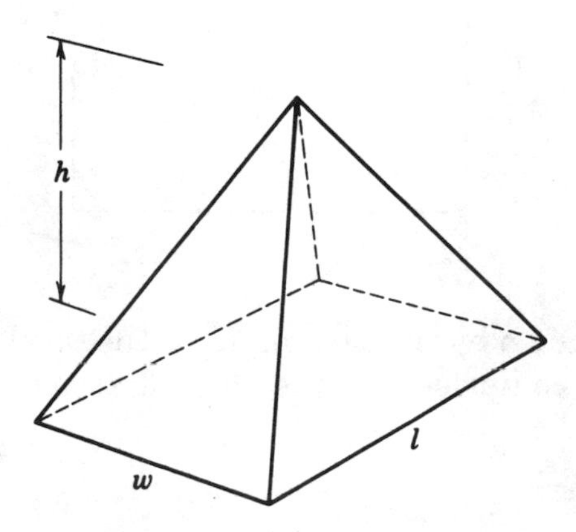

Thus,

$$dV = \frac{1}{3}\, wh\; dl + \frac{1}{3}\, lh\; dw + \frac{1}{3}\, lw\; dh.$$

If the dimensions are changed by small amounts dl, dw, and dh, the volume changes by an amount $\Delta V \approx dV$, where dV is given by the expression above.

Appendix B4
DIFFERENTIAL EQUATIONS

Any equation which involves a derivative of a function is called a *differential equation*. Such equations occur in many different applications of calculus, and their solution is the subject of a lively branch of mathematics. Here are two examples to show how a differential equation can occur.

The Growth of Population

Suppose we let n represent the number of people in a particular country. We assume that n is such a large number that we can neglect the fact that it must be an integer and treat it as a continuous positive number. (In any application we would eventually round off n to the nearest integer.) The problem is this: Assume the birthrate is proportional to the population so nA children are born every year for every n people. A is the constant of proportionality. If the initial population of the country is n_0 people, how many people are there at some later time, T? (In this simple problem we will neglect deaths.)

If there are n people, the total number of children born per year is nA. This is the *rate* of increase of population. That is,

$$\frac{dn}{dt} = nA.$$

The above differential equation is a particularly simple one. We can solve it by integration in the following manner:

$$\frac{dn}{n} = A\; dt.$$

Let us take the definite integral of both sides of the above equation. Initially we have $t = 0$ and $n = n_0$, and finally $t = T$ and $n = n(T)$. Thus

$$\int_{n_0}^{n(T)} \frac{dn}{n} = \int_0^T A\; dt.$$

The integral on the left should be familiar (if not, see Table 2, integral 5). Evaluating both integrals, we have

$$\ln n(T) - \ln n_0 = A(T - 0)$$

or

$$\ln \frac{n(T)}{n_0} = AT.$$

This equation is in the form $\ln x = AT$, where we have set $x = n(T)/n_0$. We can solve this for x by using the relation $e^{\ln x} = x$. Thus, $x = e^{\ln x} = e^{AT}$, and we have

$$\frac{n(T)}{n_0} = e^{AT}.$$

This expression describes the so-called exponential increase of population. Expressions of similar form describe many processes which are mathematically similar, for instance, the growth of money in banks due to interest or the radioactive decay of atomic nuclei.

Oscillatory Motion

As a second example of a differential equation, consider the motion of a particle in one dimension. It is sometimes possible to define the motion of the particle by a differential equation. For example, let x be the coordinate of the particle relative to the origin. Suppose we require that the position x of the particle satisfies the following differential equation:

$$\frac{d^2x}{dt^2} = -kx. \tag{1}$$

(This particular equation describes the motion of a pendulum, or of a particle suspended by a spring.)

The problem is to find out how x varies with time when it obeys this equation. This can be found by "solving" the differential equation. One of the most powerful means for solving differential equations is to guess a possible general form for the answer. Then this general form is substituted in the differential equation and one can both see that the equation is satisfied and determine any restrictions that should apply to the solution.

First, what is a promising guess as to a solution? Note that x must depend upon time in such a way that when it is differentiated twice with respect to time it reverses sign. But this is exactly what happens to the sine function since $\frac{d}{dx}(\sin x) = \cos x$, and $\frac{d^2}{dx^2}(\sin x) = \frac{d}{dx}(\cos x) = -\sin x$ (frame **212**). Therefore, let us try

$$x = A \sin(bt + c),$$

where A, b, and c are undetermined constants.

This may be differentiated twice with respect to time with the result

$$\frac{dx}{dt} = Ab \cos(bt + c),$$

$$\frac{d^2x}{dt^2} = -Ab^2 \sin(bt + c).$$

If these relations are substituted in Equation (1), we have

$$-Ab^2 \sin(bt + c) = -kA \sin(bt + c).$$

The differentiated equation is then satisfied for all t provided

$$b^2 = k.$$

(Alternatively, the equation is satisfied by $A = 0$. However, this leads to a trivial result, $x = 0$, so we disregard this possibility.) Thus the solution is

$$x = A \sin(\sqrt{k}\, t + c).$$

Although the constant k is given by Equation (1), the constants A and c are arbitrary. If the position x and the velocity dx/dt were specified at some initial time, $t = 0$, the arbitrary constants could be determined.

Note that the solution we have found corresponds to x oscillating back and forth indefinitely between $x = A$ and $x = -A$. This type of oscillatory motion is characteristic of a pendulum or of a particle suspended by a spring, so that the original differential equation really appears to describe these systems.

Appendix B5
SUGGESTIONS FOR FURTHER READING

Calculus with Analytic Geometry, 2d ed., Howard Anton, Wiley, New York, 1984.

Calculus, Dennis Berkey, W. B. Saunders & Co., Philadelphia, Pa., 1984.

Calculus and Analytic Geometry, C. H. Edwards, Jr., and David E. Penny, Prentice-Hall, Inc., Englewood Cliffs, N.J., 1982.

Calculus with Analytic Geometry, 2d ed., Robert Ellis and Denny Gulick, Harcourt Brace Jovanovich, San Diego, Calif., 1982.

Elements of Calculus and Analytic Geometry, 6th ed., Ross Finney and George Thomas, Addison-Wesley Publishing Co., Inc., Reading, Mass., 1984.

Applied Numerical Analysis, Curtis F. Gerald, Addison-Wesley Publishing Co., Inc., Reading, Mass., 1978.

Calculus, 3d ed., Stanley Grossman, Academic Press, New York, 1984.

Calculus, 2d ed., Robert E. Larson and Robert P. Hostetler, D. C. Heath, Lexington, Mass., 1981.

Calculus with Applications to Business and Life Sciences, 2d ed., Abe Mizrahi and Michael Sulliven, John Wiley & Sons, New York, 1984.

Calculus, 4th ed., Edwin J. Purcell and Dale Varberg, Prentice-Hall, Inc., Englewood Cliffs, N.J., 1984.

Calculus with Analytic Geometry, Earl Swokowski, PWS Publishers, Boston, Mass., 1982.

Review Problems

This list of problems is for benefit in case you want some additional practice. The problems are grouped according to chapter and section. Answers start on page 251.

CHAPTER 1

Linear and Quadratic Functions

Find the slope of the graphs of the following equations:

1. $y = 5x - 5$
2. $4y - 7 = 5x + 2$
3. $3y + 7x = 2y - 5$

Find the roots of the following:

4. $4x^2 - 2x - 3 = 0$
5. $x^2 - 6x + 9 = 0$

Trigonometry

6. Show that $\sin\theta \cot\theta/\sqrt{1 - \sin^2\theta} = 1$.
7. Show that $\cos\theta \sin\left(\frac{\pi}{2} + \theta\right) - \sin\theta \cos\left(\frac{\pi}{2} + \theta\right) = 1$.
8. What is: (a) $\sin 135°$, (b) $\cos \frac{7\pi}{4}$, (c) $\sin \frac{7\pi}{6}$?
9. Show that $\cos^2 \frac{\theta}{2} = \frac{1}{2}(1 + \cos\theta)$.
10. What is the cosine of the angle between any two sides of an equilateral triangle?

Exponentials and Logarithms

11. What is $(-1)^{13}$?
12. Find $[(0.01)^3]^{-1/2}$.
13. Express $\log (x^x)^x$ in terms of $\log x$.
14. If $\log(\log x) = 0$, find x.
15. Is there any number for which $x = \log x$?

In the following five questions, make use of the log table below and the rules for manipulating logarithms.

x	$\log x$	x	$\log x$
1	0.00	5	0.70
2	0.30	7	0.85
3	0.48	10	1.00

Find

16. $\log \sqrt{10}$
17. $\log 21$
18. $\log \sqrt{14}$
19. $\log 300$
20. $\log 7^{3/2}$

CHAPTER 2

Find the following limits, if they exist:

21. $\lim_{x \to 2} \frac{x^2 - 4x + 4}{x - 2}$

22. $\lim_{\theta \to \pi/2} \sin \theta$

23. $\lim_{x \to 0} \frac{x^2 + x + 1}{x}$

24. $\lim_{x \to 1} \left[1 + \frac{(x + 1)^2}{x - 1} \right]$

25. $\lim_{x \to 3} \left[(2 + x)\frac{(x - 3)^2}{x - 3} + 7 \right]$

26. $\lim_{x \to 1} \frac{x^2 - 1}{x - 1}$

27. $\lim_{x \to \infty} \frac{1}{x}$

28. $\lim_{x \to 0} \log x$

Derivatives

29. What is the average velocity of a particle that goes forward 35 miles and backward for 72 miles, during the course of 1 hour?
30. A particle always moves in one direction. Can its average velocity exceed its maximum velocity?
31. A particle moves so that its position is given by $S = S_0 \sin 2\pi t$, where S_0 is in meters, t is in hours. Find its average velocity from $t = 0$ to

    (a) $t = 1/4$ hour (b) $t = 1/2$ hour
    (c) $t = 3/4$ hour (d) $t = 1$ hour

32. Write an expression for the average velocity of a particle which leaves the origin at $t = 0$, whose position is given by $S = at^3 + bt$, where a and b are constants. The average is from $t = 0$ to the present.
33. Find the instantaneous velocity of a particle whose position is given by $S = bt^3$, where b is a constant, when $t = 2$.

Differentiation

Find the derivative of each of the following functions with respect to its appropriate variable. a and b are constants.

34. $y = x + x^2 + x^3$
35. $y = (a + bx) + (a + bx)^2 + (a + bx)^3$
36. $y = (3x^2 + 7x)^{-3}$
37. $p = \sqrt{a^2 + q^2}$
38. $p = \dfrac{1}{\sqrt{a^2 + q^2}}$
39. $y = x^{\pi}$

40. $f = \theta^2 \sin\theta$

41. $f = \dfrac{\sin\theta}{\theta}$

42. $f = (\sin\theta)^{-1}$
43. $f = (\sqrt{1 + \cos^2\theta})^{-1}$
44. $f = \sin^2\theta + \cos^2\theta$
45. $y = \sin(\ln x)$
46. $y = x \ln x$
47. $y = (\ln x)^{-2}$
48. $y = x^x$
(*Hint:* what is $\ln y$? Use implicit differentiation, Appendix B3).
49. $y = a^{x^2}$
50. $f = \sin\sqrt{1 + \theta^2}$
51. $y = e^{-x^2}$
52. $y = \pi^x$
53. $y = \pi^{x^2}$
54. $f = \ln(\sin\theta)$
55. $f = \sin(\sin\theta)$
56. $f = \ln e^x$
57. $f = e^{\ln x}$
58. $y = \sqrt{1 - \sin^2\theta}$

Higher-Order Derivatives

Evaluate each of the following:

59. Find $\dfrac{d^2}{d\theta^2}(\cos a\theta)$.

60. Find $\dfrac{d^n}{dx^n} e^{ax}$ (n is a positive integer).

61. $\dfrac{d^2}{dx^2}(\sqrt{1 + x^2})$

62. $\dfrac{d^2}{d\theta^2}(\tan\theta)$

63. $\dfrac{d^3}{dx^3}(x^2 e^x)$

Maxima and Minima

Find where the following functions have their maximum and/or minimum values. Either give the values of x explicitly, or find an equation for these values.

64. $y = e^{-x^2}$

65. $y = \dfrac{\sin x}{x}$

66. $y = e^{-x} \sin x$

67. $y = \dfrac{\ln x}{x}$

68. $y = e^{-x} \ln x$

69. Find whether y has a maximum or a minimum for the function given in question 64.

Differentials

Find the differential df of each of the following functions.

70. $f = x$
71. $f = \sqrt{x}$
72. $f = \sin x^2$
73. $f = e^{\sin x}$ (*Hint:* Use chain rule)

CHAPTER 3

You may find Table 2 on page 256 helpful in doing the problems in this section.

Integration

Find the following indefinite integrals. (Omit the constants of integration.)

74. $\int \sin 2x \, dx$

75. $\int \dfrac{dx}{x+1}$

76. $\int x^2 e^x \, dx$ (Try integration by parts.)

77. $\int x e^{-x^2} \, dx$

78. $\int \sin^2 \theta \cos \theta \, d\theta$

Some Techniques of Integration and Definite Integrals

Evaluate the following definite integrals.

79. $\int_{-1}^{+1}(e^x + e^{-x})\, dx$

80. $\int_{-\infty}^{\infty} \frac{dx}{a^2 + x^2}$

81. $\int_{-\infty}^{\infty} \frac{x\, dx}{\sqrt{a^2 + x^2}}$

82. $\int_{-\infty}^{0} x^2 e^x\, dx$ (Problem 76 may be helpful.)

83. $\int_{0}^{+\pi/2} \sin\theta \cos\theta\, d\theta$

84. $\int_{0}^{1} (x + a)^n\, dx$

85. $\int_{-1}^{+1} \frac{dx}{\sqrt{1 - x^2}}$

86. $\int_{-1}^{1} (x + x^2 + x^3)\, dx$

ANSWERS TO REVIEW PROBLEMS

1. 5
2. 5/4
3. –7
4. $(1 \pm \sqrt{13})/4$
5. 3, 3 (roots are identical)
6. No answer
7. No answer
8. (a) $\frac{\sqrt{2}}{2}$, (b) $\frac{\sqrt{2}}{2}$, (c) $-\frac{1}{2}$
9. No answer
10. ½
11. –1
12. 1000
13. $x^2 \log x$
14. $x = 10$
15. No
16. 0.50
17. 1.33
18. 0.58
19. 2.48
20. 1.28
21. 0
22. 1
23. No limit
24. No limit
25. 7
26. 2
27. 0
28. No limit
29. –37 mph
30. No
31. (a) $4S_0$ mph, (b) 0 mph, (c) $-\frac{4}{3}S_0$ mph, (d) 0 mph
32. $at^2 + b$
33. $12b$
34. $1 + 2x + 3x^2$
35. $b + 2b(a + bx) + 3b(a + bx)^2$
36. $-3(3x^2 + 7x)^{-4}(6x + 7)$
37. $\frac{dp}{dq} = \frac{q}{\sqrt{a^2 + q^2}}$

38. $\frac{dp}{dq} = \frac{-q}{(a^2 + q^2)^{3/2}}$

39. $\frac{dy}{dx} = \pi x^{\pi - 1}$

40. $\frac{df}{d\theta} = 2\theta \sin \theta + \theta^2 \cos \theta$

41. $\frac{df}{d\theta} = \frac{\cos \theta}{\theta} - \frac{\sin \theta}{\theta^2}$

42. $\frac{df}{d\theta} = -\frac{\cos \theta}{\sin^2 \theta}$

43. $\frac{df}{d\theta} = \frac{\cos \theta \sin \theta}{(1 + \cos^2 \theta)^{3/2}}$

44. $\frac{df}{d\theta} = 0$

45. $\frac{dy}{dx} = \frac{\cos(\ln x)}{x}$

46. $\frac{dy}{dx} = 1 + \ln x$

47. $\frac{dy}{dx} = \frac{-2}{x} (\ln x)^{-3}$

48. $\frac{dy}{dx} = x^x(1 + \ln x)$

49. $\frac{dy}{dx} = 2xa^{(x^2)} \ln a$

50. $\frac{\theta}{\sqrt{1 + \theta^2}} \cos \sqrt{1 + \theta^2}$

51. $-2xe^{-x^2}$
52. $\pi^x \ln \pi$
53. $2x\pi^{x^2} \ln \pi$
54. $\cot \theta$
55. $[\cos(\sin \theta)] \cos \theta$
56. 1
57. 1
58. $-\sin \theta$
59. $-a^2 \cos a\theta$
60. $a^n e^{ax}$

61. $\dfrac{1}{\sqrt{1 + x^2}} - \dfrac{x^2}{(1 + x^2)^{3/2}}$

62. $2 \sec^2 \theta \tan \theta$
63. $(6 + 6x + x^2)\, e^x$
64. $x = 0$
65. $x = \tan x \qquad (x = 0, \ldots)$
66. $x = \tan^{-1} 1 = \dfrac{\pi}{4} \pm n\pi,\ n = 0, 1, 2, \ldots$
67. $x = e \qquad (\ln x = 1)$
68. $\dfrac{1}{x} = \ln x$
69. Maximum
70. $df = dx$
71. $df = \dfrac{dx}{2\sqrt{x}}$

72. $df = 2x \cos x^2\, dx$
73. $df = \cos x e^{\sin x}\, dx$

74. $\dfrac{-1}{2} \cos 2x$

75. $\ln(x + 1)$

76. $x^2 e^x - 2xe^x + 2e^x$
77. $\dfrac{1}{2}\, e^{-x^2}$

78. $\dfrac{1}{3} \sin^3 \theta$

79. $2\left(e - \dfrac{1}{e}\right)$

80. $\dfrac{\pi}{a}$

81. 0
82. 2
83. ½

84. $\dfrac{(1 + a)^{n+1} - a^{n+1}}{n + 1}$

85. π
86. ⅔

Tables

Table 1
DERIVATIVES

The differentiation formulas are listed below. References to the appropriate frames are given. In the following expressions $\ln x$ is the natural logarithm or the logarithm to the base e; u and v are variables that depend on x; w depends on u which in turn depends on x; and a and n are constants. All angles are measured in radians.

	Formula	Frame
1.	$\frac{da}{dx} = 0$	**172**
2.	$\frac{d}{dx}(ax) = a$	**174**
3.	$\frac{dx^n}{dx} = nx^{n-1}$	**180**
4.	$\frac{d}{dx}(u + v) = \frac{du}{dx} + \frac{dv}{dx}$	**186**
5.	$\frac{d}{dx}(uv) = u\frac{dv}{dx} + v\frac{du}{dx}$	**189**
6.	$\frac{d}{dx}\left(\frac{u}{v}\right) = \frac{1}{v^2}\left(v\frac{du}{dx} - u\frac{dv}{dx}\right)$	**194**
7.	$\frac{dw}{dx} = \frac{dw}{du}\frac{du}{dx}$	**198**
8.	$\frac{du^n}{dx} = nu^{n-1}\frac{du}{dx}$	From Eqs. (3) and (7)

9.	$\dfrac{d \ln x}{dx} = \dfrac{1}{x}$	**226**
10.	$\dfrac{de^x}{dx} = e^x$	**235**
11.	$\dfrac{da^x}{dx} = a^x \ln a$	**234**
12.	$\dfrac{du^v}{dx} = vu^{v-1}\dfrac{du}{dx} + u^v \ln u \dfrac{dv}{dx}$	
13.	$\dfrac{d \sin x}{dx} = \cos x$	**211**
14.	$\dfrac{d \cos x}{dx} = -\sin x$	**212**
15.	$\dfrac{d \tan x}{dx} = \sec^2 x$	**213**
16.	$\dfrac{d \sec x}{dx} = \sec x \tan x$	**214**
17.	$\dfrac{d \cot x}{dx} = -\csc^2 x$	
18.	$\dfrac{d \sin^{-1} x}{dx} = \dfrac{1}{\sqrt{1 - x^2}}$	(Appendix B2)
19.	$\dfrac{d \cos^{-1} x}{dx} = \dfrac{-1}{\sqrt{1 - x^2}}$	(Appendix B2)
20.	$\dfrac{d \tan^{-1} x}{dx} = \dfrac{1}{1 + x^2}$	(Appendix B2)
21.	$\dfrac{d \cot^{-1} x}{dx} = \dfrac{-1}{1 + x^2}$	(Appendix B2)

Table 2
INTEGRALS

In the following u and v are variables that depend on x; w is a variable that depends on u which in turn depends on x; a and n are constants; and the arbitrary integration constants are omitted for simplicity.

1. $\int a\,dx = ax$
2. $\int af(x)\,dx = a\int f(x)\,dx$
3. $\int (u + v)\,dx = \int u\,dx + \int v\,dx$
4. $\int x^n\,dx = \dfrac{x^{n+1}}{n+1}, \qquad n \neq -1$
5. $\int \dfrac{dx}{x} = \ln x$
6. $\int e^x\,dx = e^x$
7. $\int e^{ax}\,dx = \dfrac{e^{ax}}{a}$
8. $\int b^{ax}\,dx = \dfrac{b^{ax}}{a \ln b}$
9. $\int \ln x\,dx = x \ln x - x$
10. $\int \sin x\,dx = -\cos x$
11. $\int \cos x\,dx = \sin x$
12. $\int \tan x\,dx = -\ln(\cos x)$
13. $\int \cot x\,dx = \ln(\sin x)$
14. $\int \sec x\,dx = \ln(\sec x + \tan x)$
15. $\int \sin x \cos x\,dx = \dfrac{1}{2}\sin^2 x$
16. $\int \dfrac{dx}{a^2 + x^2} = \dfrac{1}{a}\tan^{-1}\dfrac{x}{a}$
17. $\int \dfrac{dx}{\sqrt{a^2 - x^2}} = \sin^{-1}\dfrac{x}{a}$
18. $\int \dfrac{dx}{\sqrt{x^2 \pm a^2}} = \ln(x - \sqrt{x^2 \pm a^2})$
19. $\int w(u)\,dx = \int w(u)\dfrac{dx}{du}\,du$
20. $\int u\,dv = uv - \int v\,du$

Index

References are by page number. A separate index of symbols can be found on page 261.

Index of Symbols

References are by page number